Oxford Graduate

Seri

OXFORD GRADUATE TEXTS IN MATHEMATICS

Books in the series

1. Keith Hannabuss: *An Introduction to Quantum Theory*
2. Reinhold Meise and Dietmar Vogt: *Introduction to Functional Analysis*
3. James G. Oxley: *Matroid Theory*
4. N.J. Hitchin, G.B. Segal, and R.S. Ward: *Integrable Systems: Twistors, Loop Groups, and Riemann Surfaces*
5. Wulf Rossmann: *Lie Groups: An Introduction through Linear Groups*
6. Qing Liu: *Algebraic Geometry and Arithmetic Curves*
7. Martin R. Bridson and Simon M. Salamon (eds): *Invitations to Geometry and Topology*
8. Shmuel Kantorovitz: *Introduction to Modern Analysis*
9. Terry Lawson: *Topology: A Geometric Approach*
10. Meinolf Geek: *An Introduction to Algebraic Geometry and Algebraic Groups*
11. Alastair Fletcher and Vladimir Markovic: *Quasiconformal Maps and Teichmüller Theory*
12. Dominic Joyce: *Riemannian Holonomy Groups and Calibrated Geometry*
13. Fernando Villegas: *Experimental Number Theory*
14. Péter Medvegyev: *Stochastic Integration Theory*
15. Martin A. Guest: *From Quantum Cohomology to Integrable Systems*
16. Alan D. Rendall: *Partial Differential Equations in General Relativity*
17. Yves Félix, John Oprea and Daniel Tanré: *Algebraic Models in Geometry*
18. Jie Xiong: *Introduction to Stochastic Filtering Theory*
19. Maciej Dunajski: *Solitons, Instantons, and Twistors*
20. Graham R. Allan: *Introduction to Banach Spaces and Algebras*
21. James Oxley: *Matroid Theory, Second Edition*
22. Simon Donaldson: *Riemann Surfaces*
23. Clifford Henry Taubes: *Differential Geometry: Bundles, Connections, Metrics and Curvature*

Preface

This book is based on lectures given by the authors at an instructional conference on integrable systems held at the Mathematical Institute in Oxford in September 1997. Most of the participants were graduate students from the United Kingdom and other European countries. The lectures emphasized geometric aspects of the theory of integrable systems, particularly connections with algebraic geometry, twistor theory, loop groups, and the Grassmannian picture.

We are grateful for support for the conference from the London Mathematical Society, the Engineering and Physical Sciences Research Council (contract No. 00985SCI96), the University of Oxford Mathematical Prizes Fund, the Mathematical Institute, Wadham College, and Oxford University Press.

N. M. J. Woodhouse
Oxford, February 1998

Contents

Contributors

N. J. Hitchin The Mathematical Institute, University of Oxford, 24–29 St Giles', Oxford OX1 3LB.

G. B. Segal Department of Pure Mathematics and Statistics, University of Cambridge, 16 Mill Lane, Cambridge CB2 1SB.

R. S. Ward Department of Mathematical Sciences, University of Durham, South Road, Durham DH1 3LE.

1
Introduction

Nigel Hitchin

Integrable systems, what are they? It's not easy to answer precisely. The question can occupy a whole book (Zakharov 1991), or be dismissed as Louis Armstrong is reputed to have done once when asked what jazz was—'If you gotta ask, you'll never know!'

If we steer a course between these two extremes, we can say that integrability of a system of differential equations should manifest itself through some generally recognizable features:

- the existence of many conserved quantities;
- the presence of algebraic geometry;
- the ability to give explicit solutions.

These guidelines should be interpreted in a very broad sense: the algebraic geometry is often transcendental in nature, and explicitness *doesn't* mean solvability in terms of sines, exponentials or rational functions.

The most classical example of integrable systems shows all these properties: the motion of a rigid body about its centre of mass. If $\mathbf{\Omega}$ is the angular velocity vector in the body and I_1, I_2, I_3 the principal moments of inertia, then these equations take the form

$$\begin{aligned} I_1\dot{\Omega}_1 &= (I_2 - I_3)\Omega_2\Omega_3 \\ I_2\dot{\Omega}_2 &= (I_3 - I_1)\Omega_3\Omega_1 \\ I_3\dot{\Omega}_3 &= (I_1 - I_2)\Omega_1\Omega_2 \,. \end{aligned}$$

To analyse them it is easier to rescale and obtain the simpler equations

$$\begin{aligned} \dot{u}_1 &= u_2u_3 \\ \dot{u}_2 &= u_3u_1 \\ \dot{u}_3 &= u_1u_2 \,. \end{aligned}$$

So what do we look for first? *Conserved quantities.* Note that differentiating $u_1^2 - u_2^2$ gives $2u_1(u_2u_3) - 2u_2(u_3u_1) = 0$ and so $u_1^2 - u_2^2$ is constant. We similarly get

$$\begin{aligned} u_1^2 - u_2^2 &= A \\ u_1^2 - u_3^2 &= B\,. \end{aligned}$$

So A and B are two conserved quantities as (u_1, u_2, u_3) evolves.

What about *algebraic geometry*? Take the first equation $\dot{u}_1 = u_2u_3$ and substitute for u_2 and u_3 given by the expressions above, then we obtain

$$\dot{u}_1^2 = (u_1^2 - A)(u_1^2 - B)\,.$$

Putting $y = \dot{u}_1$ and $x = u_1$, we can rewrite this as

$$y^2 = (x^2 - A)(x^2 - B)$$

which is the equation of an algebraic curve, in fact an elliptic curve, and

$$\mathrm{d}t = \mathrm{d}x/y$$

is a regular differential form on the curve.

Finally how about *explicit solutions*? Any elliptic curve can be written in a standard form

$$y^2 = 4x^3 - g_2x - g_3$$

and there is a meromorphic function, the Weierstrass $\wp$-function, which is doubly periodic:

$$\wp(u + 2m\omega_1 + 2n\omega_2) = \wp(u)$$

and satisfies

$$\wp(u)'^2 = 4\wp(u)^3 - g_2\wp(u) - g_3\,.$$

Using the $\wp$-function, the solution becomes

$$\mathrm{d}t = \mathrm{d}\wp/\wp' = \mathrm{d}u$$

This means not only that if we are prepared to use elliptic functions, we can solve the equation, but also that time in the original equation is linear in the natural parameter u: we have achieved in some sense a linearization of the non-linear differential equation for the rigid body.

The study of integrable systems is not just about cunning methods of solving isolated special equations. Each equation is slightly different, and indeed there are many of them: a trawl through a couple of standard books on the subject gives at least the following list of equations which are seriously considered to be related to integrability:

Calogero-Moser system, Calogero–Sutherland system, Euler–Arnold rigid body, Clebsch rigid body, Euler–Poinsot top, Garnier system, Gaudin system, Goryachev–Chaplyagin top, Henon–Heiles system, Kepler problem, Kirchoff rigid body, Kowalewski top, Lagrange top, Neumann problem,Toda lattice, Ruijsenaars system, Steklov rigid body, Nahm's equation, Boussinesq equation, Burger's equation, Davey–Stewartson equation, Drinfeld–Sokolov construction, Ernst equation, Painlevé equation, Euler–Arnold–Manakov equation, Gelfand–Levitan–Marchenko equation, Heisenberg ferromagnet equation, Korteweg–de Vries equation, Kadomtsev–Pietviashvili equation, Krichever construction, Landau–Lifschitz equation, Hasimoto equation, Lax equation, Liouville equation, Manakov–Zakharov model, modified KdV equation, nonlinear Schrödinger equation, Riccati equation, Schlesinger equation, sine–Gordon equation, Zakharov–Shabat equation, Benjamin–Ono equation, Calogero–Degasperis–Fokas equation, Harry–Dym equation, Fermi–Pasta–Ulam problem, massive Thirring model, Melnikov equation, Benjamin–Bona–Mahoney equation, Maxwell–Bloch equation ...,

Another task of the mathematician, apart from solving individual equations, is to put some order into a universe like this. Is there some overarching structure of which all these are special cases which explains integrability?

The point where most discussions of integrability begin is with the idea of a system of differential equations which can be put in *Lax pair* form. Let's begin with a finite-dimensional system

$$\frac{\mathrm{d}A}{\mathrm{d}t} = [A, B]$$

where

$$A(\mathbf{z}) = A_0 + \mathbf{z}A_1 + \cdots + \mathbf{z}^n A_n \qquad B(\mathbf{z}) = B_0 + \mathbf{z}B_1 + \cdots + \mathbf{z}^m B_m$$

are polynomials of $k \times k$ matrices. Because of the differential equation, we have

$$\begin{aligned}\frac{\mathrm{d}}{\mathrm{d}t}\operatorname{tr}(A^p) &= \operatorname{tr}(p[A,B]A^{p-1}) \\ &= p\operatorname{tr}(ABA^{p-1} - BA^p) \\ &= p\operatorname{tr}(BA^p - BA^p) \\ &= 0\end{aligned}$$

so all the coefficients of the polynomials $\operatorname{tr} A(\mathbf{z})^p$ for all p are conserved quantities. Since the components of the characteristic polynomial are expressible in terms of these traces, it is the whole spectrum of $A(\mathbf{z})$ which is preserved. Clearly equations of Lax pair type satisfy the first criterion for integrability that there should be many conserved quantities. In fact, algebraic geometry appears again very naturally.

The characteristic equation

$$\det(\mathbf{y} - A(\mathbf{z})) = 0$$

defines an algebraic curve, called the spectral curve, which is preserved by the flow. For each point $(\mathbf{y}, \mathbf{z})$ on this curve we have a one-dimensional space

$$L_{(\mathbf{y},\mathbf{z})} = \ker(\mathbf{y} - A(\mathbf{z}))$$

and this varies with time—it forms a line bundle over the spectral curve. To study equations of this type, then one must study the algebraic geometry of algebraic curves and line bundles over them.

It is a well-known fact that the space of line bundles is a complex torus—the quotient space of a vector space $\mathbb{C}^g$ by a lattice subgroup, and it is here that the final criterion of explicitness of solutions is fulfilled. We regard the equation as integrable if the line bundle L moves in a linear fashion with t in this vector space. Under these circumstances the solutions can in principle be written down in terms of theta-functions. It requires a specific form for the matrix $B(\mathbf{z})$ to be able to do this, however. An arbitrary matrix B would give a non-linear isospectral deformation of $A(\mathbf{z})$.

Many finite-dimensional integrable systems fit into this scheme, and in particular the rigid body, where we can take

$$\begin{aligned} A(\mathbf{z}) &= \begin{pmatrix} 0 & (u_1+u_2) \\ (u_2-u_1) & 0 \end{pmatrix} + \mathbf{z}\begin{pmatrix} -2u_3 & 0 \\ 0 & 2u_3 \end{pmatrix} \\ &\quad + \mathbf{z}^2\begin{pmatrix} 0 & (u_1-u_2) \\ -(u_1+u_2) & 0 \end{pmatrix}. \end{aligned}$$

Finite-dimensional integrable systems are rather special, but are a good model for the general situation without having to worry too much about analytical problems. In infinite dimensions we can still make use of the Lax pair formalism, though. We have an equation

$$\frac{\mathrm{d}A}{\mathrm{d}t} = [A, B]$$

but now A and B are differential operators instead of finite-dimensional matrices. The most famous example here is that of the Korteweg–de Vries equation

$$u_t - 6uu_x + u_{xxx} = 0$$

where

$$A = -\frac{\mathrm{d}^2}{\mathrm{d}x^2} + u \qquad B = 4\frac{\mathrm{d}^3}{\mathrm{d}x^3} - 3u\frac{\mathrm{d}}{\mathrm{d}x} - 3\frac{\mathrm{d}}{\mathrm{d}x}u\,.$$

Here, as $u(x,t)$ evolves, the spectrum of A remains fixed, but there are different types of spectrum depending on the boundary conditions, and the role of algebraic geometry, which in the finite-dimensional situation manifests itself as the spectral curve, has to be extended.

The most algebraic aspect is in the case of periodic boundary conditions. The spectrum of the basic operator $-\mathrm{d}^2/\mathrm{d}x^2$ is n^2, $n \in \mathbb{Z}$, and the multiplicity of each eigenvalue is two unless $n = 0$. In general, the simple spectrum may be finite or infinite. For the finite case, there is a close analogy with the finite-dimensional situation of a 2×2 matrix $A(\mathbf{z})$ of polynomials. In that case the spectral curve

$$\det(\mathbf{y} - A(\mathbf{z})) = \mathbf{y}^2 + a_1(\mathbf{z})y + a_2(\mathbf{z}) = 0$$

can be described as a double covering of the line branched over the zeros of $a_1^2 - 4a_2$. In the case that the differential operator A has a finite simple spectrum, the abstract curve, or Riemann surface, formed by taking such a double covering branched over the simple spectrum plays the same role, and indeed the same hyperelliptic theta functions can be used to solve the KdV equation as in the finite-dimensional case. This, however, is only a special case even within the class of periodic boundary conditions. When the simple spectrum is infinite, the algebraic geometry must be modified to describe hyperelliptic Riemann surfaces of infinite genus.

The case where $u(x,t)$ is a Schwartz function on the line involves the consideration of both the discrete spectrum and the continuous spectrum. The role of a line bundle on a spectral curve is now played by the asymptotic behaviour of the eigenfunctions, the reflection coefficients, and recovering the potential $u(x,t)$ is part of a process called the Gelfand–Levitan–Marchenko inverse scattering transform. One defines a function

$$B(x) = \sum_{n=1}^{N} c_n^2 \mathrm{e}^{-z_n x} + \frac{1}{2\pi}\int_{-\infty}^{\infty} b(z) e^{\mathrm{i}xz}\,\mathrm{d}z$$

based on the spectrum and reflection coefficients, solves an integral equation

$$B(x+z) + \int_{-\infty}^{\infty} K(x,y)B(y+z)\,\mathrm{d}y = -K(x,z)$$

and then finds the function $u(x)$ by:

$$u(x) = -2\frac{\mathrm{d}}{\mathrm{d}x}K(x,x)\,.$$

Now let's move on to the situation where the operator A in the Lax pair

$$\frac{\mathrm{d}A}{\mathrm{d}t} = [A,B]$$

is defined by a first-order matrix differential operator instead of a higher-order scalar equation

$$A(\mathbf{z}) = \frac{\mathrm{d}}{\mathrm{d}x} + C(\mathbf{z}),$$

depending on a parameter $\mathbf{z}$, which is the analogue of the spectral parameter for the scalar differential operator. We can again look at different boundary conditions and spectral problems, but we may also reformulate the Lax pair equations in a different language: that of differential geometry. Firstly, we write the equations as

$$\left[\frac{\partial}{\partial x} + C, \frac{\partial}{\partial t} + B\right] = 0$$

and then, putting

$$\nabla_0 = \frac{\partial}{\partial x} + C, \qquad \nabla_1 = \frac{\partial}{\partial t} + B,$$

we recognize the equation

$$[\nabla_0, \nabla_1] = 0$$

as the equation for the vanishing of the curvature of a connection, with covariant derivatives ∇_0, ∇_1, on a bundle over two-dimensional space. More correctly, we have a family of connections parametrized by $\mathbf{z}$, all of which are flat.

Many standard equations fit into this zero curvature formalism, for example the nonlinear Schrödinger equation:

$$q_t = -\frac{\mathrm{i}}{2}(q_{xx} + 2|q|^2 q)$$

where

$$C(\mathbf{z}) = \begin{pmatrix} -\mathrm{i}\mathbf{z} & q \\ -\bar{q} & \mathrm{i}\mathbf{z} \end{pmatrix}, \qquad B(\mathbf{z}) = \mathbf{z}C(\mathbf{z}) + \begin{pmatrix} \mathrm{i}|q|^2/2 & \mathrm{i}q_x/2 \\ -\mathrm{i}\bar{q}_x/2 & -\mathrm{i}|q|^2/2 \end{pmatrix}$$

and the familiar KdV equation again

$$q_t = -\tfrac{1}{4}(6qq_x + q_{xxx})$$

where

$$\begin{aligned} C(\mathbf{z}) &= \begin{pmatrix} -\mathrm{i}\mathbf{z} & q \\ -1 & \mathrm{i}\mathbf{z} \end{pmatrix}, \\ B(\mathbf{z}) &= \mathbf{z}^2 C(\mathbf{z}) + \mathbf{z}\begin{pmatrix} \mathrm{i}q/2 & \mathrm{i}q_x/2 \\ 0 & -\mathrm{i}q/2 \end{pmatrix} + \begin{pmatrix} q_x/2 & -q^2/2 \\ q/2 & -q_x/4 \end{pmatrix}. \end{aligned}$$

The zero curvature condition is a convenient entry into one of the most promising unifying ideas in integrable systems: trying to view known systems as special cases of the anti-self-dual Yang–Mills equations (Mason and Woodhouse 1996). These gauge-theoretic equations in four dimensions have become of great importance in many disciplines. For the purposes of studying integrable systems, one starts with four-dimensional space with a flat metric of indefinite signature:

$$ds^2 = 2(\mathrm{d}t\mathrm{d}u - \mathrm{d}v\mathrm{d}w)$$

and a connection on a bundle over this space with covariant derivatives

$$\nabla_t = \frac{\partial}{\partial t} + A_t, \qquad \text{etc.}$$

The anti-self-dual Yang–Mills equations are then defined as

$$[\nabla_t, \nabla_v] = 0 \qquad [\nabla_u, \nabla_w] = 0$$

$$[\nabla_t, \nabla_u] - [\nabla_v, \nabla_w] = 0$$

and these can be encapsulated in a single equation with a parameter $\mathbf{z}$:

$$[\nabla_v - \mathbf{z}\nabla_u, \nabla_t - \mathbf{z}\nabla_w] = 0 \,.$$

Geometrically, this is a zero curvature condition: through each point $\mathbb{R}^4$ there are two families of totally null 2-planes, α-planes and β-planes, and the anti-self-dual equations say that the connection has zero curvature on each α-plane.

The integrable systems which arise from this approach are obtained by dimensional reduction: looking at solutions invariant under some group of conformal transformations. The anti-self-duality condition then leads to equations on some lower-dimensional space. The standard examples of evolution equations arise from considering invariance under the additive group

$$(t, u, v, w) \mapsto (t, u + a, v + b, w - b) \,.$$

We have invariant functions $v + w = x$ (=space) and t (=time) and the metric induces a degenerate metric $\mathrm{d}x^2$ on this two-dimensional space. After using gauge transformations and symmetries of $\mathrm{d}x^2$ (Galilean transformations), some remarkable features appear when finding canonical forms. For example, for connection matrices which are 2×2 and trace-free, there are effectively only two reduced forms: one gives the nonlinear Schrödinger equation, the other the KdV equation.

The use of the formalism of connections, curvature and gauge transformations effects a geometrization of problems concerning integrable systems. One can still usually detect the key ingredients, but in a different language. For example, scattering data is usually seen as the holonomy of a flat connection, and the algebro-geometrical content as a manifestation of Penrose's twistor theory: the well-known result of Ward and Penrose that the anti-self-dual Yang–Mills equations can be encoded in the geometry of a holomorphic vector bundle over $\mathbb{P}^3$.

While the use of the anti-self-dual Yang–Mills equations as a unifying tool in integrable systems is impressive, there is no complete agreement about whether these are truly universal equations. The KP equation, for example, needs to be ruthlessly hacked and stretched to fit the Procrustean bed of self-duality.

The equations which appear in integrable systems very often have a specific mechanical or physical source: the shallow water waves of the KdV equation, the exponential attraction of particles for the Toda lattice, non-linear optics for the non-linear Schrödinger equation, and so forth. It is not profitable, however, to dwell too long on any one particular physical source since the equations in some sense go deeper. The physical problem can only be approximated by the equations, but the equations are highly special and do not retain their features after perturbation. Moreover, the same equations can arise from many different sources. As an example, consider the non-linear Schrödinger equation again. Far from non-linear optics and gauge theory, it essentially arose in the study of vortex filaments in the paper of Da Rios (1906). This can be experienced with some elementary differential geometry.

Take a closed curve $\mathbf{r}(s)$ in $\mathbb{R}^3$ (the 'vortex') and let it evolve in time along its *binormal* according to the equation

$$\frac{\partial \mathbf{r}}{\partial t} = \kappa \mathbf{b}$$

where κ is the curvature. This is tantalizingly close to the more standard mean curvature flow, where $\mathbf{b}$ is replaced by $\mathbf{n}$, the normal, but the solutions are very different. Whereas the mean curvature flow decreases the length as fast as possible, the length is preserved under the binormal flow. This is but one of many conserved quantities, and in fact Hasimoto showed in 1972 that putting

$$q(s) = \kappa(s) \exp \left(\mathrm{i} \int^{s} \kappa(u) \, \mathrm{d}u \right)$$

the function q satisfies the non-linear Schrödinger equation. One may then apply to this simple geometrical problem the full force of the solution techniques for integrable systems to obtain explicit solutions of the evolution of these curves. For examples and pictures of this and other geometrical problems solved by integrable system methods, the reader is referred to Melko and Sterling (1993).

In so many areas, from physics to geometry, one encounters (sometimes in hidden form) the phenomenon of integrability. Recognizing it, and becoming acquainted with the techniques for exploiting it, has become a necessity for a broad band of mathematicians.

Bibliography

Da Rios, L. S. (1906). Sul moto d'un liquido indefinito con un filetto vorticoso di forma qualinque. *Rend. Circ. Mat. Palermo*, **22**.

Mason, L. J. and Woodhouse, N. M. J. (1996). *Integrability, self-duality, and twistor theory.* Oxford University Press.

Melko, M. and Sterling, I. (1993). Integrable systems, harmonic maps and the classical theory of solitons. In *Harmonic maps and integrable systems*, (ed. A. P. Fordy and J. C. Wood). Vieweg, Wiesbaden.

Zakharov, V. E. (1991). *What is integrability?* Springer series in non-linear dynamics. Springer-Verlag, Berlin.

2
Riemann surfaces and integrable systems

Nigel Hitchin

Notes by Justin Sawon

1 Riemann surfaces

In this chapter, we shall investigate the solution of certain types of integrable systems by studying line bundles on Riemann surfaces. In particular, we shall be interested in integrable systems of finite-dimensional type given by Lax pair equations

$$\frac{\mathrm{d}A}{\mathrm{d}t} = [A, B]$$

where

$$A = A_0 + A_1 z + \cdots + A_m z^m$$

and

$$B = B_0 + B_1 z + \cdots + B_n z^n$$

are polynomial-valued matrices. Expanding out the time-evolution equation in different powers of z we get a system of differential equations in the coefficients A_i and B_j of A and B.

It might seem that this gives an enormous number of equations and the systems which occur in real life cannot possibly be of this form, but in fact many of them are, but for very specific forms of matrices. These include classical problems like the geodesic flow on an ellipsoid, or the motion of a rigid body, but also more recently studied ones like Nahm's equations, which are dimensional reductions of the self-dual Yang–Mills equations and occur in the analysis of magnetic monopoles. We shall cover these only in examples, however, and focus on the solution of the general equation of the above type.

To do so requires using the basic language of Riemann surfaces and the first two sections will be taken up with a rapid treatment of the subject. In lectures three and four we shall look at the perhaps less standard material of vector bundles on Riemann surfaces. The fifth lecture is on Lax pairs and their relation to line bundles. In the final lecture we shall look at algebraically completely integrable Hamiltonian systems—a general context of which many of these Lax pair equations form part.

General references for the theory of Riemann surfaces are Gunning (1966, 1967, 1972), Farkas and Kra (1980), and Griffiths and Harris (1978) for a more general account of algebraic geometry. The material on Lax pairs is dealt with in various places in the mathematical literature, but an accessible introduction is Audin's book (Audin 1996).

Definition 1.1 *A Riemann surface is a one-dimensional complex manifold, i.e. a two-real-dimensional smooth manifold with a maximal set of coordinate charts $\phi_\alpha : U_\alpha \to \mathbb{R}^2 = \mathbb{C}$ such that $\phi_\beta \circ \phi_\alpha^{-1}$ is an invertible holomorphic function from $\phi_\alpha(U_\alpha \cap U_\beta)$ to $\phi_\beta(U_\alpha \cap U_\beta)$ for all α and β.*

Thus a neighbourhood of any point can be parametrized by a complex number z and on an overlapping neighbourhood with parameter w, $w(z)$ is a holomorphic function of one variable.

Examples

(1) Let $M = S^2$, the sphere. Stereographic projection from the North pole (denoted N) to the tangent plane at the South pole S defines a coordinate chart $\phi_0 : S^2\backslash\{N\} \to \mathbb{C}$. Similarly, stereographic projection from the South pole defines a coordinate chart $\phi_1 : S^2\backslash\{S\} \to \mathbb{C}$. A calculation shows that, using the correct orientations on the two tangent planes, $\phi_1 \circ \phi_0^{-1}(z) = z^{-1}$, which is clearly holomorphic as a function from $\mathbb{C}^*$ to $\mathbb{C}^*$.

With this complex structure, the two-sphere is known as the Riemann sphere: the parameter z of the first chart runs through $\mathbb{C}$ and then we can take N to be the point $z = \infty$. It is also known as the projective line $\mathbb{P}^1$, i.e. the space of one-dimensional subspaces of $\mathbb{C}^2$. We shall keep this as the standard notation.

(2) Let $M = \mathbb{C}^*/\mathbb{Z}$, where the integer n acts by $z \mapsto \lambda^n z$, with λ a complex number with absolute value not equal to one. (Without loss of generality, assume that $|\lambda| > 1$.) Topologically M is a torus, T^2. We can take overlapping annuli in $\mathbb{C}^*$ as coordinate neighbourhoods.

Definition 1.2 *A holomorphic map $f : \tilde{M} \to M$ is a continuous map such that for each coordinate chart $\phi_\alpha : U_\alpha \to \mathbb{C}$ on M and $\tilde{\phi}_\beta : \tilde{U}_\beta \to \mathbb{C}$ on $\tilde{M}$, $\phi_\alpha \circ f \circ \tilde{\phi}_\beta^{-1}$ is holomorphic.*

Examples

(1) Let $M = \mathbb{P}^1$ and $\tilde{M} = \mathbb{P}^1$. We shall use the coordinate z corresponding to the coordinate chart $\tilde{\phi}_0$. Then the rational function

$$f(z) = \frac{p(z)}{q(z)} = \frac{a_0 + a_1 z + \cdots + a_k z^k}{b_0 + b_1 z + \cdots + b_l z^l}$$

defines a holomorphic map from $\mathbb{P}^1$ to itself, where we have assumed that p and q have no common zeros. The point is that where $f(z)$ is infinite, $1/f(z)$ is holomorphic, so in either coordinate on the target space $\phi_\alpha \circ f \circ \tilde{\phi}_0^{-1}$ is holomorphic. The same is true for $\tilde{\phi}_1$ since $f(1/z)$ is a rational function too. In fact, all holomorphic maps from $\mathbb{P}^1$ to itself are given by rational functions.

(2) In general, a meromorphic function on M can be interpreted as a holomorphic map of Riemann surfaces $f : M \to \mathbb{P}^1$. Where the meromorphic function acquires a pole the point is mapped to the North pole.

(3) Let $M = \mathbb{C}^*/\mathbb{Z}$, and define f by the sum

$$f(z) = \sum_{n=-\infty}^{\infty} \frac{z}{(\lambda^{n/2} z - \lambda^{-n/2})^2},$$

for $z \in \mathbb{C}^*$. This sum converges for $1 \leq |z| \leq |\lambda|$ and $z \neq \lambda^k$, for integers k. Furthermore,

$$\begin{aligned} f(\lambda z) &= \sum \frac{\lambda z}{(\lambda^{n/2} \lambda z - \lambda^{-n/2})^2} \\ &= \sum \frac{z}{(\lambda^{(n+1)/2} z - \lambda^{-(n+1)/2})^2} \\ &= f(z), \end{aligned}$$

so f is invariant under the action of the integers. Therefore we have a holomorphic map $f : M \to \mathbb{P}^1$.

Consider a holomorphic map $f : M \to \mathbb{C}$, usually referred to as a holomorphic function on M.

Theorem 1.3 *If M is connected and compact, the only holomorphic functions on M are the constants.*

This follows from the maximum modulus principle: $|f|$ has to have a maximum on the compact space M, but in a coordinate neighbourhood of this point $f \circ \phi_\alpha^{-1}$ is a holomorphic function whose modulus has an interior maximum. Despite the theorem, there are many *meromorphic* functions on any compact Riemann surface as we shall see. It is more useful to view this in the context of holomorphic line bundles on Riemann surfaces.

Definition 1.4 *A holomorphic line bundle L over a Riemann surface M is a two-dimensional complex manifold L with a holomorphic projection $\pi : L \to M$ such that*

(1) *for each $m \in M$, $\pi^{-1}(m)$ has the structure of a one-dimensional vector space,*

(2) *each point $m \in M$ has a neighbourhood U and a homeomorphism φ_U such that*

$$\begin{array}{ccc} \pi^{-1}(U) & \overset{\varphi_U}{\cong} & U \times \mathbb{C} \\ & {\scriptstyle \pi} \searrow \quad \swarrow & \\ & U & \end{array}$$

is commutative,

(3) *$\varphi_V \circ \varphi_U^{-1}$ is of the form*

$$(m, w) \mapsto (m, f(m)w),$$

where f is a non-vanishing holomorphic function.

We will denote f by g_{VU}. These functions are known as the transition functions of the line bundle and φ_U is a local trivialization over U.

In straightforward language, a holomorphic line bundle is a family of one-dimensional vector spaces parametrized by M.

Examples

(1) $M \times \mathbb{C}$ is known as the trivial bundle over M.

(2) Take a point $p \in M$, and U_0 a neighbourhood of p with coordinate z such that $z(p) = 0$. Let $U_1 = M \backslash \{p\}$. Then we can use z as a transition function to define a line bundle on M since $g_{01} = z$ is holomorphic and non-vanishing on $U_0 \cap U_1$. We patch together $U_0 \times \mathbb{C}$ and $U_1 \times \mathbb{C}$ over $U_0 \cap U_1$ by using the function φ defined by

$$\varphi(m, w) = (m, g_{01}(m)w)\,.$$

This gives us for each point $p \in M$ a line bundle which we denote by L_p.

Definition 1.5 *A holomorphic section of a line bundle L over M is a holomorphic map $s : M \to L$ such that $\pi \circ s = \mathrm{id}_M$.*

In a local trivialization φ_U of the line bundle, the section gives

$$\varphi_U(s) = (\mathrm{id}, s_U) : U \to U \times \mathbb{C}$$

and is therefore just defined by a holomorphic function s_U on U. On the overlap $U \cap V$, these local functions are related by

$$s_U = g_{UV} s_V,$$

and we can think of a section s as a collection of local functions $\{s_U\}$ that patch together in this way.

If s and t are two holomorphic sections of L, then

$$s_U = g_{UV} s_V$$

and

$$t_U = g_{UV} t_V$$

together imply that

$$\frac{s_U}{t_U} = \frac{s_V}{t_V}$$

on the overlap $U \cap V$. Therefore we can piece these local meromorphic functions together to get a global meromorphic function on M. If we find lots of sections of line bundles, we can then find lots of meromorphic functions in this way.

We can also add sections pointwise

$$(s + t)(m) := s(m) + t(m)$$

and multiply sections by scalars

$$(\lambda s)(m) := \lambda s(m),$$

so that the space of all sections of L is a vector space, which we denote $\mathrm{H}^0(M, L)$.

Theorem 1.6 *If M is compact, $\mathrm{H}^0(M, L)$ is finite-dimensional.*

The proof of this result can be found in, for example, Gunning (1966). From the point of view of analysis it is an example of the fact that a linear elliptic differential operator on a compact manifold has a finite-dimensional kernel.

Examples

(1) The line bundle L_p has a canonical section s_p: we just take the two functions z on U_0 and 1 on U_1 and this constitutes a section since

$$z = z.1 = g_{01}1\,.$$

The section s_p has a simple zero at p and only there. We shall use this many times in forthcoming sections.

(2) The *canonical bundle* K is the cotangent bundle, or the bundle of holomorphic 1-forms. Suppose we have local coordinates z and w, with $w(z) = \phi_\beta \circ \phi_\alpha^{-1}(z)$ a function of z on the overlap. The 1-forms $\mathrm{d}z$ and $\mathrm{d}w$ give local trivializations of the canonical bundle, and on the overlap $\mathrm{d}w = w'\mathrm{d}z$. Therefore the transition functions are $\mathrm{d}w/\mathrm{d}z$, where $w = \phi_\beta \circ \phi_\alpha^{-1}$.

(3) Consider $M = \mathbb{P}^1$ with the usual coordinate patches U_0 and U_1. The transition function $g_{01} = z^n$ on $U_0 \cap U_1 \cong \mathbb{C}^*$ defines a line bundle which we usually denote by $\mathcal{O}(n)$. A section of this line bundle is given by functions s_0 and s_1 on $\mathbb{C}$ related by

$$s_0(z) = z^n s_1(\tilde{z})$$

on the overlap $\mathbb{C}^*$. Expanding these functions as power series in their respective local coordinates, and using the fact that $\tilde{z} = z^{-1}$, we get

$$\sum_0^\infty a_m z^m = z^n \sum_0^\infty \tilde{a}_m z^{-m}\,.$$

Equating coefficients, we find that $\tilde{a}_m = a_m = 0$ for $m > n$ and $\tilde{a_0} = a_n$, $\tilde{a_1} = a_{n-1}$, etc. Thus the section is given by a polynomial

$$\sum_0^n a_m z^m$$

of degree less than or equal to n, and hence the dimension of

$$\mathrm{H}^0(\mathbb{P}^1, \mathcal{O}(n))$$

is $n+1$.

One of our aims in this chapter is to construct polynomial-valued matrices

$$A = A_0 + A_1 z + \cdots + A_m z^m$$

which occur in the Lax pair equations. Instead of thinking in terms of polynomials we can now interpret $A(z)$ more geometrically as a matrix with values in the space of sections $\mathrm{H}^0(\mathbb{P}^1, \mathcal{O}(m))$.

2 Line bundles and sheaves

Any natural operation that can be carried out on one-dimensional vector spaces transfers to an operation on line bundles, and given one or two we can construct many. Here are the essential ones, and the corresponding operation on transition functions:

Properties of line bundles

(1) Given L we can form its dual bundle L^*, also denoted L^{-1}. It has transition functions $g_{\alpha\beta}(L^*) = g_{\alpha\beta}^{-1}(L)$.

(2) Given L and $\tilde{L}$ we can form their tensor product $L \otimes \tilde{L}$. It has transition functions $g_{\alpha\beta}(L \otimes \tilde{L}) = g_{\alpha\beta}(L) g_{\alpha\beta}(\tilde{L})$.

(3) We can also form the homomorphism bundle $\mathrm{Hom}(L, \tilde{L}) \cong L^* \otimes \tilde{L}$. Holomorphic homomorphisms between line bundles are just holomorphic sections of this.

(4) The bundle of endomorphisms $\mathrm{Hom}(L, L) \cong L^* \otimes L$ is canonically trivial because the only endomorphisms of a one-dimensional vector space are the scalars. This explains why we write L^* also as L^{-1}.

(5) If s is a section of L and $\tilde{s}$ a section of $\tilde{L}$, then the product $s\tilde{s}$ is a section of $L \otimes \tilde{L}$, usually denoted by just $L\tilde{L}$.

(6) If $f : \tilde{M} \to M$ is a holomorphic map, then we define the pull-back of line bundle L on M by

$$f^*L := \{(x, q) \in L \times \tilde{M} : \pi(x) = f(q)\}\,.$$

Its transition functions are $g_{\alpha\beta} \circ f$. A section of f^*L is a holomorphic map $s : \tilde{M} \to L$ such that $\pi \circ s = f$.

Let us consider a little closer now the canonical bundle K. Any Riemann surface has a canonical bundle, and its vector space of sections is an important invariant.

Definition 2.1 *If M is compact, its genus g is defined to be the dimension of* $\mathrm{H}^0(M, K)$.

Examples

(1) Let $M = \mathbb{P}^1$. A section of the canonical bundle looks like $f_0(z)\,\mathrm{d}z$ on U_0 and $f_1(\tilde{z})\,\mathrm{d}\tilde{z}$ on U_1 where f_0 and f_1 are holomorphic functions on $\mathbb{C}$. These 1-forms must agree on the overlap $U_0 \cap U_1$. Here we have $\tilde{z} = z^{-1}$, and therefore

$$\mathrm{d}\tilde{z} = -z^{-2}\,\mathrm{d}z\,.$$

It follows that we must have

$$f_0(z)\,\mathrm{d}z = -z^{-2} f_1(z^{-1})\,\mathrm{d}z,$$

and expanding f_0 and f_1 as power series shows that they both must vanish. Therefore there are no non-zero global sections of the canonical bundle, and the genus of $\mathbb{P}^1$ is zero. Note that $d\tilde{z}$ and $-dz$ are related by the transition function z^{-2}, so on $\mathbb{P}^1$ we have an isomorphism $K \cong \mathcal{O}(-2)$.

(2) Let $M = \mathbb{C}^*/\mathbb{Z}$. Then $\mathrm{d}z/z$ defines a 1-form on $\mathbb{C}^*$, and since

$$\frac{\mathrm{d}(\lambda z)}{(\lambda z)} = \frac{\mathrm{d}z}{z}$$

it is invariant under the action of the integers. Therefore this 1-form defines a non-vanishing section of K. In general, if a line bundle L has a non-vanishing section s, then

$$\begin{array}{ccc} M \times \mathbb{C} & \to & L \\ (m, u) & \mapsto & us(m) \end{array}$$

is an isomorphism between L and the trivial bundle. Thus for $M = \mathbb{C}^*/\mathbb{Z}$ the canonical bundle is trivial. Sections of the trivial bundle are just functions and since the only holomorphic functions are constants, the genus of $\mathbb{C}^*/\mathbb{Z}$ is one.

One of the consequences of using concepts which are locally standard, like coordinates on a manifold, or local trivializations of a line bundle, is that we end up with familiar objects defined on open sets and their intersections. Thus sections of line bundles are given by functions f_α on U_α and line bundles themselves by transition functions $g_{\alpha\beta}$ on $U_\alpha \cap U_\beta$. The technology to handle globally such notions is encapsulated in sheaf theory and its cohomology. The definition of a sheaf seeks to formalize the properties of restricting functions from open sets:

Definition 2.2 *A sheaf $\mathcal{S}$ on a topological space X associates to each open set $U \subset X$ an abelian group $\mathcal{S}(U)$ (sections over U) and to $U \subset V$ a restriction map $r_{VU} : \mathcal{S}(V) \to \mathcal{S}(U)$ such that*

(1) *for $U \subset V \subset W$, $r_{WU} = r_{VU} \circ r_{WV}$;*
(2) *if $\sigma \in \mathcal{S}(U)$ and $\tau \in \mathcal{S}(V)$, and $r_{UU\cap V}(\sigma) = r_{VU\cap V}(\tau)$ then there exists $\rho \in \mathcal{S}(U \cup V)$ such that $r_{U\cup VU}(\rho) = \sigma$ and $r_{U\cup VV}(\rho) = \tau$;*
(3) *if $\sigma \in \mathcal{S}(U \cup V)$ is such that $r_{U\cup VU}(\sigma) = 0$ and $r_{U\cup VV}(\sigma) = 0$ then $\sigma = 0$.*

Examples

(1) $\mathcal{S}(U) = \mathcal{O}(U) =$ holomorphic functions on U.

(2) $\mathcal{S}(U) = \mathcal{O}(L)(U) =$ sections of the holomorphic line bundle L over U.

(3) $\mathcal{S}(U) =$ locally constant functions on U with values in $\mathbb{C}$ or $\mathbb{Z}$.

(4) $\mathcal{S}(U) = \mathcal{O}^*(U) =$ non-vanishing holomorphic functions on U, with the group operation being multiplication.

If $\mathcal{S}$ is a sheaf, we can construct the cohomology groups $\mathrm{H}^p(M, \mathcal{S})$ with coefficients in $\mathcal{S}$ in the following way. Take a (locally finite) covering $\{U_\alpha\}_{\alpha \in A}$ of M by open sets. Let

$$S^0 = \bigoplus_\alpha \mathcal{S}(U_\alpha),$$

$$S^1 = \bigoplus_{\alpha \neq \beta} \mathcal{S}(U_\alpha \cap U_\beta),$$

$$S^p = \bigoplus_{\alpha_0 \neq \cdots \neq \alpha_p} \mathcal{S}(U_{\alpha_0} \cap \cdots \cap U_{\alpha_p}).$$

and let C^p be the alternating elements in S^p. In other words, for a permutation of the indices $\alpha_0, \ldots, \alpha_p$, the open set is unchanged but we multiply the section on that set by the sign of the permutation.

Define the homomorphism of abelian groups $\delta : C^p \to C^{p+1}$ by

$$(\delta f)_{\alpha_0 \ldots \alpha_{p+1}} = \sum_i (-1)^i f_{\alpha_0 \cdots \hat{\alpha}_i \ldots \alpha_{p+1}} |_{U_{\alpha_0} \cap \ldots \cap U_{\alpha_{p+1}}} .$$

It is known as the boundary operator, and satisfies $\delta^2 = 0$. Thus we can define the following quotient groups.

Definition 2.3 *The p-th cohomology group of $\mathcal{S}$, relative to this covering, is*

$$\mathrm{H}^p(M, \mathcal{S}) := \frac{\ker \delta : C^p \to C^{p+1}}{\operatorname{im} \delta : C^{p-1} \to C^p} .$$

Remark The dependence of these cohomology groups on the covering might be a cause for the reader's concern. The standard method to make a cohomology group independent of the covering is to take a limit over *all* coverings partially ordered by refinement. What one gets then is clearly independent but is not calculable. The way out is to use a 'good' covering: one for which the cohomology of all the intersections vanishes for $p > 0$. For such a covering the sheaf cohomology is the same as the limit.

Examples

(1) Let L be a holomorphic line bundle over M and $\mathcal{S}$ the sheaf of holomorphic sections. If $f \in C^0$, then $(\delta f)_{\alpha\beta} = f_\alpha - f_\beta$. Thus δf vanishes if and only if the local sections f_α piece together to give a global section, i.e.

$$\mathrm{H}^0(M, L) = \ker \delta$$

is the space of global holomorphic sections of L (which agrees with our earlier notation).

(2) Suppose the line bundle L has transition functions $g_{\alpha\beta}$ defined by $\varphi_\alpha \circ \varphi_\beta^{-1}$, where $\varphi_\alpha : \pi^{-1}(U_\alpha) \to U_\alpha \times \mathbb{C}$ are the local trivializations of the bundle. Then $g_{\alpha\beta} = g_{\beta\alpha}^{-1}$ and so lies in C^1 for the sheaf $\mathcal{O}^*$ of non-vanishing holomorphic functions, and

$$\begin{aligned}(\delta g)_{\alpha\beta\gamma} &= g_{\alpha\beta} g_{\beta\gamma} g_{\alpha\gamma}^{-1} \\ &= \varphi_\alpha \varphi_\beta^{-1} \varphi_\beta \varphi_\gamma^{-1} (\varphi_\alpha \varphi_\gamma^{-1})^{-1} \\ &= \mathrm{id}\,,\end{aligned}$$

i.e. $\delta g = 1$ (note that the group operation is multiplicative in this example). The φ_α are not unique. If we change the local trivialization φ_α to $h_\alpha \varphi_\alpha$, then g changes to $g(\delta h)$. By the same token, the transition functions of two isomorphic bundles differ by δh for some $h \in C^1$ and it follows that the isomorphism classes of holomorphic line bundles on a Riemann surface (or complex manifold in general) are given by elements of the sheaf cohomology group $\mathrm{H}^1(M, \mathcal{O}^*)$.

Although the definition of sheaf cohomology groups may seem very complicated, in fact for Riemann surfaces, we only need $p = 0, 1$ and occasionally $p = 2$, so that we can be very concrete in describing classes. In particular, we have the following theorem (see Gunning 1966 for a proof).

Theorem 2.4 *Let M be a Riemann surface. If $\mathcal{S} = \mathcal{O}(L)$, the sheaf of holomorphic sections of the line bundle L, then $\mathrm{H}^p(M, \mathcal{S}) = 0$ for $p > 1$. If $\mathcal{S} = \mathbb{C}$ or $\mathbb{Z}$, then $\mathrm{H}^p(M, \mathcal{S}) = 0$ for $p > 2$.*

It is also true that $\mathrm{H}^1(M, \mathcal{O}(L))$ (which we shall now write simply as $\mathrm{H}^1(M, L)$) can be defined in terms of the more familiar space of holomorphic sections of a line bundle. This is

Theorem 2.5. (Serre duality) *If L is a line bundle on a compact Riemann surface M, then*

$$\mathrm{H}^1(M, L) \cong \mathrm{H}^0(M, K \otimes L^*)^*\,.$$

Again, see Gunning (1966) for a proof.

With sheaves, we can define subsheaves and quotient sheaves. (The latter is a little subtle to define in general but will be obvious in all our examples.) In particular if $\mathcal{S}$ is a subsheaf of $\mathcal{T}$ with quotient $\mathcal{U}$ then there is a highly important relationship between their cohomology groups which we shall use over and over again:

Theorem 2.6 *If*

$$0 \to \mathcal{S} \to \mathcal{T} \to \mathcal{U} \to 0$$

is a short exact sequence of sheaves on M, then there is a long exact sequence of cohomology groups

$$0 \to \mathrm{H}^0(M,\mathcal{S}) \to \mathrm{H}^0(M,\mathcal{T}) \to \mathrm{H}^0(M,\mathcal{U}) \xrightarrow{\delta_0} \mathrm{H}^1(M,\mathcal{S}) \to \cdots.$$

$$\cdots \to \mathrm{H}^p(M,\mathcal{S}) \to \mathrm{H}^p(M,\mathcal{T}) \to \mathrm{H}^p(M,\mathcal{U}) \xrightarrow{\delta_p} \mathrm{H}^{p+1}(M,\mathcal{S}) \to \cdots$$

We can describe the coboundary operator δ_0 in the following way. Suppose that $\{u_\alpha\} \in \mathrm{H}^0(M,\mathcal{U})$; then it satisfies $u_\alpha - u_\beta = 0$. There exists $\{t_\alpha\} \in C^0(\mathcal{T})$, not necessarily uniquely defined, such that $t_\alpha \mapsto u_\alpha$. Now $\{t_\alpha - t_\beta\} \in C^1(\mathcal{T})$ maps to $u_\alpha - u_\beta = 0$, so by exactness of the short exact sequence there exists a unique $s_{\alpha\beta} \in C^1(\mathcal{S})$ such that $s_{\alpha\beta} \mapsto t_\alpha - t_\beta$. It is easily shown that $\delta s = 0$, and hence $s \in \mathrm{H}^1(M,\mathcal{S})$. Then we define $\delta_0 u := s$.

Example Let L be a line bundle on M, and L_p the line bundle associated to a point $p \in M$. Recall that L_p has a section s_p which vanishes only at p. There is a short exact sequence

$$0 \to \mathcal{O}(LL_p^{-1}) \xrightarrow{s_p} \mathcal{O}(L) \to \mathcal{O}_p(L) \to 0,$$

where $\mathcal{O}_p(L)(U)$ can be interpreted as the sections of L over $U \cap \{p\}$. It has a one-dimensional space of global sections which is simply the vector space $\pi^{-1}(p)$. This gives rise to the long exact sequence

$$0 \to \mathrm{H}^0(M,LL_p^{-1}) \to \mathrm{H}^0(M,L) \to \mathbb{C} \xrightarrow{\delta} \mathrm{H}^1(M,LL_p^{-1}) \to \cdots.$$

If δ is non-zero, then the map from $\mathrm{H}^0(M,L)$ to $\mathbb{C}$ must be zero, and by exactness we have an isomorphism

$$\mathrm{H}^0(M,LL_p^{-1}) \cong \mathrm{H}^0(M,L)$$

given by multiplication by the section s_p. However, s_p vanishes at p, and so it follows that if δ is non-zero then all global sections of L must vanish at p.

3 Vector bundles

Ultimately, we are going to produce matrix polynomials $A(z)$ from line bundles, and consider time evolution of such matrices. Since these describe paths in the space of equivalence classes of line bundles we would like to understand that space, namely $\mathrm{H}^1(M, \mathcal{O}^*)$, better. It is a good opportunity to learn the use of the sheaf cohomological results of the last section.

We begin by considering the short exact sequence of sheaves

$$0 \to \mathbb{Z} \to \mathcal{O} \overset{\exp(2\pi \mathrm{i} f)}{\longrightarrow} \mathcal{O}^* \to 1\,.$$

This gives rise to a long exact sequence

$$0 \to \mathbb{Z} \to \mathbb{C} \to \mathbb{C}^* \to \mathrm{H}^1(M,\mathbb{Z}) \to \mathrm{H}^1(M,\mathcal{O}) \to \mathrm{H}^1(M,\mathcal{O}^*) \to$$
$$\to \mathrm{H}^2(M,\mathbb{Z}) \to \mathrm{H}^2(M,\mathcal{O}) \to \cdots\,.$$

The first part of this sequence comes from the fact that holomorphic functions on compact Riemann surfaces are constants. Since exponentiation is surjective onto $\mathbb{C}^*$ then from exactness $\mathrm{H}^1(M,\mathbb{Z})$ injects into $\mathrm{H}^1(M,\mathcal{O})$. We also know that $\mathrm{H}^2(M,\mathcal{O})$ must vanish, so this sequence reduces to

$$0 \to \frac{\mathrm{H}^1(M,\mathcal{O})}{\mathrm{H}^1(M,\mathbb{Z})} \to \mathrm{H}^1(M,\mathcal{O}^*) \overset{\delta}{\to} \mathrm{H}^2(M,\mathbb{Z}) \to 0\,.$$

Topological considerations (M is compact of real dimension two) tell us that $\mathrm{H}^2(M,\mathbb{Z}) \cong \mathbb{Z}$.

Definition 3.1 *The degree of a line bundle L is $\delta([L])$. It is denoted* $\deg L$ *(or $c_1(L)$, as it is also the first Chern class).*

Properties

(1) If L and $\tilde{L}$ are two line bundles, then

$$\deg(L \otimes \tilde{L}) = \deg L + \deg \tilde{L}\,.$$

This is just the fact that δ is a homomorphism.

(2) If L_p is the line bundle corresponding to the point p, then $\deg L_p = 1$. This is essentially a normalization. In topological terms it comes from the fact that the generator of $\mathrm{H}^2(M,\mathbb{Z}) \cong \mathbb{Z}$ comes from the generator of $\mathrm{H}^1(S^1,\mathbb{Z}) \cong \mathbb{Z}$ in a Mayer–Vietoris sequence.

(3) If a section $s \in \mathrm{H}^0(M, L)$ vanishes at the points $p_1, \ldots, p_n$ with multiplicities $m_1, \ldots, m_n$ then $\deg L = \sum_i m_i$.

To prove the last of these properties, we notice that $ss_{p_1}^{-1}$ gives a section of $LL_{p_1}^{-1}$ which vanishes at p_1 with multiplicity $m_1 - 1$. Thus $ss_{p_1}^{-m_1} \ldots s_{p_n}^{-m_n}$ gives a non-vanishing section of $LL_{p_1}^{-m_1} \ldots L_{p_n}^{-m_n}$, and hence $LL_{p_1}^{-m_1} \ldots L_{p_n}^{-m_n}$ is trivial. The trivial bundle clearly has degree zero, and thus using properties 1 and 2, we see that $\deg L = \sum_i m_i$. An obvious corollary is:

Corollary *If* $\deg L < 0$, *then* L *has no non-trivial holomorphic sections.*

We return now to our investigation of $\mathrm{H}^1(M, \mathcal{O}^*)$. By Serre duality,

$$\mathrm{H}^1(M, \mathcal{O}) \cong \mathrm{H}^0(M, K)^*$$

and so is g-dimensional. Consider the short exact sequence

$$0 \to \mathbb{C} \to \mathcal{O} \xrightarrow{d} \mathcal{O}(K) \to 0,$$

where d is the derivative of functions. This gives rise to a long exact sequence

$$0 \to \mathbb{C} \xrightarrow{\cong} \mathbb{C} \to \mathrm{H}^0(M, \mathcal{O}(K)) \to \mathrm{H}^1(M, \mathbb{C}) \to \mathrm{H}^1(M, \mathcal{O}) \to$$

$$\to \mathrm{H}^1(M, \mathcal{O}(K)) \to \mathrm{H}^2(M, \mathbb{C}) \to 0\,.$$

By Serre duality,

$$\mathrm{H}^1(M, \mathcal{O}(K)) \cong \mathrm{H}^0(M, \mathcal{O})^* \cong \mathbb{C},$$

and topological considerations tell us that $\mathrm{H}^2(M, \mathbb{C}) \cong \mathbb{C}$ also. Thus the map from $\mathrm{H}^1(M, \mathcal{O}(K))$ to $\mathrm{H}^2(M, \mathbb{C})$ is an isomorphism, and hence the map from $\mathrm{H}^1(M, \mathcal{O})$ to $\mathrm{H}^1(M, \mathcal{O}(K))$ must be zero. Both $\mathrm{H}^0(M, \mathcal{O}(K))$ and $\mathrm{H}^1(M, \mathcal{O})$ are g-dimensional, and it follows by exactness that

$$\dim \mathrm{H}^1(M, \mathbb{C}) = 2g\,.$$

We can also say that

$$\mathrm{H}^1(M, \mathbb{Z}) = \mathbb{Z}^{2g},$$

as there is no torsion in H^1. Therefore our exact sequence for $\mathrm{H}^1(M,\mathcal{O}^*)$ becomes

$$0 \to \frac{\mathbb{C}^g}{\mathbb{Z}^{2g}} \to \mathrm{H}^1(M,\mathcal{O}^*) \to \mathbb{Z} \to 0\,.$$

The group $\mathrm{H}^1(M,\mathcal{O}^*)$ is known as the Picard group of M. The first term in the sequence is simply a quotient of $\mathbb{C}^g$ by a lattice, which is topologically a $2g$-dimensional torus.

Thus each line bundle has an integer invariant, its degree, and the space of equivalence classes of line bundles of given degree d (which we shall denote by J^d) is a complex torus. These tori are all isomorphic to the *Jacobian* of the Riemann surface, and we shall use this word henceforth. We call a straight line in the Jacobian the image of a straight line in $\mathbb{C}^g$. Later we will exploit this when we look at integrable systems; time evolution will become evolution of the line bundle, which will be a straight line evolution, thereby linearizing the non-linear problem.

If the classification of line bundles on Riemann surfaces is essentially linear the same is not true of vector bundles, which also play a role in integrable systems.

Definition 3.2 *A rank m vector bundle over a Riemann surface M is a complex manifold E with a holomorphic projection $\pi : E \to M$ such that*

(1) *for each $z \in M$, $\pi^{-1}(z)$ is an m-dimensional complex vector space,*

(2) *each point $z \in M$ has a neighbourhood U and a homeomorphism φ_U such that*

$$\begin{array}{ccc} \pi^{-1}(U) & \overset{\varphi_U}{\cong} & U \times \mathbb{C}^m \\ & {\scriptstyle\pi} \searrow \quad \swarrow & \\ & U & \end{array}$$

is commutative,

(3) *$\varphi_V \circ \varphi_U^{-1}$ is of the form*

$$(z, w) \mapsto (z, A(z)w),$$

where $A : U \cap V \to \mathrm{GL}(m,\mathbb{C})$ is a holomorphic map to the space of invertible $m \times m$ matrices.

Remark As with line bundles, we denote A by g_{VU}. These transition functions still satisfy $g_{UV}g_{VW} = g_{UW}$. However, they are matrix-valued, and so do not commute in general. This means, for example, that sheaf theory cannot be used to classify vector bundles in the same way as for line bundles—it is very difficult to try and adapt sheaf theory to non-abelian groups.

General constructions

(1) Given two vector bundles E and $\tilde{E}$, we can form their direct sum $E \oplus \tilde{E}$, and $\mathrm{rk}(E \oplus \tilde{E}) = \mathrm{rk}E + \mathrm{rk}\tilde{E}$ (we use $\mathrm{rk}E$ to denote the rank of E).

(2) We can also form their tensor product, $E \otimes \tilde{E}$.

(3) We can take the dual bundle, E^*.

(4) The highest exterior power forms a line bundle $\det(E) = \bigwedge^m E$. More concretely, this is the line bundle with transition functions $\det(g_{\alpha\beta})$.

Definition 3.3 *If E is a vector bundle, we define its degree by*

$$\deg(E) := \deg(\det(E))\,.$$

This is the same as the first Chern class of E, $c_1(E)$.

Let $\mathcal{O}(E)$ be the sheaf of holomorphic sections of the vector bundle E (defined in the same way as for a line bundle). Then just as for line bundles,

$$\mathrm{H}^p(M, \mathcal{O}(E)) = 0$$

for $p > 1$. Moreover, Serre duality also holds for vector bundles.

For vector bundles, as with line bundles, the vector spaces

$$\mathrm{H}^p(M, \mathcal{O}(E))$$

are important objects. There is a fundamental theorem, which we shall prove next, which relates their dimensions:

Theorem 3.4. (Riemann–Roch) *If E is a vector bundle on a compact Riemann surface of genus g, then*

$$\dim \mathrm{H}^0(M, E) - \dim \mathrm{H}^1(M, E) = \deg E + \mathrm{rk}\, E(1 - g)\,.$$

Proof We shall prove this result by induction on the rank of E.

(1) Assume the rank of E is one, i.e. $E = L$ is a line bundle.

(a) If L is the trivial bundle then $\mathcal{O}(L) = \mathcal{O}$, and

$$\dim \mathrm{H}^0(M, \mathcal{O}) - \dim \mathrm{H}^1(M, \mathcal{O}) = 1 - g$$

and

$$\deg \mathcal{O} + \operatorname{rk} \mathcal{O}(1-g) = 0 + 1(1-g),$$

so the formula holds.

(b) We claim that if the formula holds for a line bundle L, then it will also hold for LL_p^{-1} and LL_p. First consider the short exact sequence

$$0 \to \mathcal{O}(L) \xrightarrow{s_p} \mathcal{O}(LL_p) \to \mathcal{O}_p(LL_p) \to 0\,.$$

The corresponding long exact sequence is

$$0 \to \mathrm{H}^0(M, L) \to \mathrm{H}^0(M, LL_p) \to \mathbb{C} \to \mathrm{H}^1(M, L) \to \mathrm{H}^1(M, LL_p) \to 0\,.$$

Exactness tells us that the alternating sum of the dimensions in this series must be zero, and therefore

$$\begin{aligned}
\dim \mathrm{H}^0(M, LL_p) - \dim \mathrm{H}^1(M, LL_p) & \\
&= \dim \mathrm{H}^0(M, L) - \dim \mathrm{H}^1(M, L) + 1 \\
&= \deg L + (1-g) + 1 \\
&= \deg(LL_p) + (1-g).
\end{aligned}$$

This proves the formula for LL_p; the proof for LL_p^{-1} is similar.

(c) We now show that every line bundle L is isomorphic to some product

$$L_{p_1} \ldots L_{p_m} L_{q_1}^{-1} \ldots L_{q_n}^{-1}\,.$$

Consider the short exact sequence

$$0 \to \mathcal{O}(L) \xrightarrow{s_p^n} \mathcal{O}(LL_p^n) \to \mathcal{S} \to 0,$$

where the quotient sheaf $\mathcal{S}$ is described locally as the quotient space of $f(z) \mapsto z^n f(z)$ where z is a coordinate vanishing at p. This quotient thus captures the first n coefficients of the Taylor expansion of $f(z)$ at p. The corresponding long exact sequence is

$$0 \to \mathrm{H}^0(M, L) \to \mathrm{H}^0(M, LL_p^n) \to \mathbb{C}^n \to \mathrm{H}^1(M, L) \to \mathrm{H}^1(M, LL_p^n) \to 0\,.$$

Using the fact that the alternating sum of the dimensions of the terms in a long exact sequence must vanish, we find that

$$\begin{aligned} &\dim\ \mathrm{H}^0(M, LL_p^n) \\ &\quad = \ n + \dim \mathrm{H}^1(M, LL_p^n) + \dim \mathrm{H}^0(M, L) - \dim \mathrm{H}^1(M, L) \\ &\quad \geq \ n + \dim \mathrm{H}^0(M, L) - \dim \mathrm{H}^1(M, L). \end{aligned}$$

Choosing n sufficiently large, the right-hand side is positive, which means there exists a holomorphic section s of LL_p^n. Suppose this section vanishes at $p_1, \ldots, p_k$ with multiplicities $m_1, \ldots, m_k$. Then $ss_{p_1}^{-m_1} \ldots s_{p_k}^{-m_k}$ is a non-vanishing section which trivializes the bundle $LL_p^n L_{p_1}^{-m_1} \ldots L_{p_k}^{-m_k}$. Therefore we get an isomorphism

$$L \cong L_{p_1}^{m_1} \ldots L_{p_k}^{m_k} L_p^{-n},$$

as required, and it follows that the formula holds for all line bundles.

(2) Suppose that E is a vector bundle of rank m, and we assume inductively that the theorem holds for bundles of lower rank. We shall find a line bundle as a subbundle of E. Note that a line subbundle $L \subset E$ is the same as a non-vanishing section of $\mathrm{Hom}(L, E) = L^* \otimes E$. Consider the short exact sequence

$$0 \to \mathcal{O}(E) \stackrel{s_p^n}{\to} \mathcal{O}(E \otimes L_p^n) \to \mathcal{S} \to 0\,.$$

As before, we get

$$\dim \mathrm{H}^0(M, E \otimes L_p^n) \geq mn + \dim \mathrm{H}^0(M, E) - \dim \mathrm{H}^1(M, E),$$

so for large enough n we can find a section s of $E \otimes L_p^n$. Suppose that s vanishes at $p_1, \ldots, p_k$ with multiplicities $m_1, \ldots, m_k$. Then $ss_{p_1}^{-m_1} \ldots s_{p_k}^{-m_k}$ is a non-vanishing section of $E \otimes L^*$, where $L^* = L_p^n L_{p_1}^{-m_1} \ldots L_{p_k}^{-m_k}$. Thus we have an inclusion $L \subset E$. Now consider the short exact sequence

$$0 \to \mathcal{O}(L) \to \mathcal{O}(E) \to \mathcal{O}(Q) \to 0,$$

where Q is the rank $m-1$ quotient bundle. Using the fact that the alternating sum of the dimensions in the corresponding long exact sequence is zero, and using our inductive hypothesis, we get

$$\begin{aligned}
&\dim \mathrm{H}^0(M,E) - \dim \mathrm{H}^1(M,E) \\
&\quad = \dim \mathrm{H}^0(M,L) - \dim \mathrm{H}^1(M,L) + \dim \mathrm{H}^0(M,Q) \\
&\qquad - \dim \mathrm{H}^1(M,Q) \\
&\quad = \deg L + (1-g) + \deg Q + (m-1)(1-g) \\
&\quad = \deg E + m(1-g),
\end{aligned}$$

as $\deg E = \deg L + \deg Q$. This follows from the fact that $\det E = L \otimes \det Q$. Thus the formula holds for E, which concludes the proof. □

4 Direct images of line bundles

In the previous sections we have looked at line bundles and vector bundles over arbitrary Riemann surfaces M. Now we specialize to the Riemann sphere $M = \mathbb{P}^1$, which has genus 0, and ask: what are the holomorphic vector bundles on $\mathbb{P}^1$?

First begin with line bundles. Recall that we constructed $\mathcal{O}(n)$ on $\mathbb{P}^1$ by using the transition function z^n on $U_0 \cap U_1$, where $\mathbb{P}^1 = U_0 \cup U_1$ is the standard covering. We also saw that line bundles are classified up to isomorphism by the Picard group $\mathrm{H}^1(M, \mathcal{O}^*)$, which fits into the exact sequence

$$0 \to \frac{\mathrm{H}^1(M,\mathcal{O})}{\mathrm{H}^1(M,\mathbb{Z})} \to \mathrm{H}^1(M,\mathcal{O}^*) \xrightarrow{\deg} \mathbb{Z} \to 0\,.$$

The first term has the structure of a g complex-dimensional torus. Since $g = 0$ in this case, this term must vanish. Thus the Picard group is isomorphic to the integers, and so the degree classifies line bundles on $\mathbb{P}^1$ up to holomorphic isomorphism. For example, if $p \in \mathbb{P}^1$ then L_p has degree one, and so must be isomorphic to $\mathcal{O}(1)$, which is L_0 as the transition function z vanishes at $z = 0$. Similarly, $L_{p_1} \dots L_{p_m} \cong \mathcal{O}(m)$.

The classification of holomorphic vector bundles is equally simple to state:

Theorem 4.1. (Birkhoff–Grothendieck) *If E is a rank m holomorphic vector bundle over $\mathbb{P}^1$, then*

$$E \cong \mathcal{O}(a_1) \oplus \cdots \oplus \mathcal{O}(a_m)$$

for some $a_i \in \mathbb{Z}$.

Proof We will prove this theorem by induction on the rank of E. We have already seen that it is true for line bundles, so suppose E is a rank m vector bundle. We have seen previously that for n sufficiently large, $E(n) = E \otimes \mathcal{O}(n)$ will have holomorphic sections. Consider the short exact sequence

$$0 \to \mathcal{O}E(n-1) \stackrel{s_p}{\to} \mathcal{O}E(n) \to \mathcal{S} \to 0,$$

where $\mathcal{S}$ is the quotient sheaf. We can deduce from the corresponding long exact sequence that the induced map

$$\mathrm{H}^0(\mathbb{P}^1, E(n-1)) \stackrel{s_p}{\to} \mathrm{H}^0(\mathbb{P}^1, E(n))$$

is injective. Suppose these groups have the same dimension; then the above map must be an isomorphism, which implies that all sections of $E(n)$ must vanish at p. Since this is true for all points $p \in \mathbb{P}^1$, we have a contradiction. Therefore

$$\dim \mathrm{H}^0(\mathbb{P}^1, E(n-1)) < \dim \mathrm{H}^0(\mathbb{P}^1, E(n)),$$

and so there exists an integer n such that

$$\mathrm{H}^0(\mathbb{P}^1, E(n-1)) = 0$$

and

$$\mathrm{H}^0(\mathbb{P}^1, E(n)) \neq 0 .$$

The long exact sequence now looks like

$$0 \to 0 \to \mathrm{H}^0(\mathbb{P}^1, E(n)) \to \mathrm{H}^0(\mathbb{P}^1, \mathcal{S}) \to \mathrm{H}^1(\mathbb{P}^1, E(n-1)) \to \cdots .$$

If s is a non-trivial section of $E(n)$, then the map to $\mathrm{H}^0(\mathbb{P}^1, \mathcal{S})$ is given by evaluation at the point p. By exactness, this map is injective, and hence $s(p) \neq 0$. Since this is true for all $p \in \mathbb{P}^1$, s is a non-vanishing section. Therefore it defines an inclusion of the trivial line bundle $\mathcal{O}$ in $E(n)$ by

$$\begin{array}{rccc} \mathcal{O} = & M \times \mathbb{C} & \to & E(n) \\ & (m, \lambda) & \mapsto & \lambda s(m). \end{array}$$

So we have an exact sequence

$$0 \to \mathcal{O} \to E(n) \xrightarrow{\alpha} Q \to 0,$$

where Q is the quotient bundle. For $\mathcal{O}$ to split off from $E(n)$ in a direct sum decomposition, we require a copy of Q inside $E(n)$ which is complementary to $\mathcal{O}$. This is a splitting of the exact sequence, i.e. a homomorphism $Q \to E(n)$ which gives the identity on Q when composed with α. To show that one exists, consider the short exact sequence obtained from the above by tensoring with Q^*

$$0 \to Q^* = \mathrm{Hom}(Q, \mathcal{O}) \to \mathrm{Hom}(Q, E(n)) \to \mathrm{Hom}(Q, Q) \to 0,$$

and the corresponding long exact sequence

$$0 \to \mathrm{H}^0(\mathbb{P}^1, Q^*) \to \mathrm{H}^0(\mathbb{P}^1, \mathrm{Hom}(Q, E(n))) \to$$

$$\to \mathrm{H}^0(\mathbb{P}^1, \mathrm{Hom}(Q, Q)) \to \mathrm{H}^1(\mathbb{P}^1, Q^*) \to \cdots .$$

Clearly there is a non-vanishing section of $\mathrm{Hom}(Q, Q)$ given by the identity map I from Q to Q. We would like to show that I maps to zero in $\mathrm{H}^1(\mathbb{P}^1, Q^*)$, as this would mean it lifts to a section of $\mathrm{Hom}(Q, E(n))$, which is what we want.

By our inductive hypothesis, Q splits into a direct sum of line bundles

$$Q = \mathcal{O}(b_1) \oplus \cdots \oplus \mathcal{O}(b_{m-1}) .$$

Consider the short exact sequence

$$0 \to \mathcal{O}(-1) \to \mathcal{O}E(n-1) \to \mathcal{O}Q(-1) \to 0$$

and the corresponding long exact sequence

$$0 \to \mathrm{H}^0(\mathbb{P}^1, \mathcal{O}(-1)) \to \mathrm{H}^0(\mathbb{P}^1, E(n-1)) \to$$

$$\to \mathrm{H}^0(\mathbb{P}^1, Q(-1)) \to \mathrm{H}^1(\mathbb{P}^1, \mathcal{O}(-1)) \to \cdots .$$

The first of these groups vanishes as $\mathcal{O}(-1)$ has negative degree, the second vanishes due to the way we chose n. By the Riemann–Roch theorem applied to $\mathcal{O}(-1)$, we see that

$$\begin{aligned}\dim \mathrm{H}^1(\mathbb{P}^1, \mathcal{O}(-1)) &= \dim \mathrm{H}^0(\mathbb{P}^1, \mathcal{O}(-1)) - \deg \mathcal{O}(-1) - (1-g)\\ &= 0 - (-1) - (1-0)\\ &= 0\,,\end{aligned}$$

and so the fourth group vanishes also. Therefore

$$\mathrm{H}^0(\mathbb{P}^1, Q(-1)) = \bigoplus_i \mathrm{H}^0(\mathbb{P}^1, \mathcal{O}(b_i - 1)) = 0\,,$$

and it follows that $b_i - 1$ must be negative for all i, since for $n \geq 0$, $\mathcal{O}(n)$ has an $(n+1)$-dimensional space of sections. Thus $b_i \leq 0$. Now applying the Riemann–Roch theorem to $\mathcal{O}(-b_i)$, we see that

$$\begin{aligned}\dim \mathrm{H}^1(\mathbb{P}^1, \mathcal{O}(-b_i)) &= \dim \mathrm{H}^0(\mathbb{P}^1, \mathcal{O}(-b_i)) - \deg \mathcal{O}(-b_i)\\ &\quad - (1-g)\\ &= (-b_i + 1) - (-b_i) - (1-0)\\ &= 0\,,\end{aligned}$$

as the sections of $\mathcal{O}(-b_i)$ are polynomials of degree $-b_i$. It follows that

$$\mathrm{H}^1(\mathbb{P}^1, Q^*) = \bigoplus_i \mathrm{H}^1(\mathbb{P}^1, \mathcal{O}(-b_i)) = 0\,.$$

Therefore I lifts to $\mathrm{H}^0(\mathbb{P}^1, \mathrm{Hom}(Q, E(n)))$, or in other words, there exists a section of $\mathrm{Hom}(Q, E(n))$, which means that $E(n)$ splits as $\mathcal{O} \oplus Q$. Thus

$$E \cong \mathcal{O}(-n) \oplus \mathcal{O}(b_1 - n) \oplus \cdots \oplus \mathcal{O}(b_{m-1} - n),$$

which concludes the proof. □

Corollary *Let E be a holomorphic vector bundle over $\mathbb{P}^1$. Then E is trivial if and only if* $\deg E = 0$ *and* $\mathrm{H}^0(\mathbb{P}^1, E(-1)) = 0$.

Proof The only if statement is obvious. Conversely, suppose E has degree zero and satisfies $\mathrm{H}^0(\mathbb{P}^1, E(-1)) = 0$. By the Birkhoff–Grothendieck theorem, we can assume

$$E = \mathcal{O}(a_1) \oplus \cdots \oplus \mathcal{O}(a_m)$$

for some $a_i \in \mathbb{Z}$. Then

$$\deg E = \sum_i a_i = 0,$$

and

$$\mathrm{H}^0(\mathbb{P}^1, E(-1)) = \bigoplus_i \mathrm{H}^0(\mathbb{P}^1, \mathcal{O}(a_i - 1)) = 0$$

implies that $a_i - 1$ is negative for all i, i.e. $a_i \leq 0$. It follows that $a_i = 0$ for all i, and hence E is trivial. □

We are now going to produce vector bundles on M from line bundles on a covering in a natural way. Suppose we have a holomorphic map $f : \tilde{M} \to M$ between compact Riemann surfaces. The degree of f is defined to be

$$\deg f := \deg(f^* L_p),$$

for $p \in M$. Since L_p has a holomorphic section that vanishes at p, its pull-back vanishes with total multiplicity $\deg f$ and this, if p is a regular value, is just the number of points in $f^{-1}(p)$.

Given a sheaf $\mathcal{S}$ on $\tilde{M}$, we define in a canonical way the *direct image sheaf* $f_*\mathcal{S}$ on M by

$$(f_*\mathcal{S})(U) := \mathcal{S}(f^{-1}(U))\,.$$

Proposition 4.2 *Take $\mathcal{S} = \mathcal{O}(L)$ for some line bundle L on $\tilde{M}$. Then*

(1) $\mathrm{H}^0(M, f_*\mathcal{O}(L)) \cong \mathrm{H}^0(\tilde{M}, \mathcal{O}(L))$,

(2) $f_*\mathcal{O}(L) = \mathcal{O}(E)$ *on M, for E a rank m holomorphic vector bundle, where $m = \deg f$,*

(3) *if V is a holomorphic vector bundle on M, then*

$$f_*\mathcal{O}(L \otimes f^*V) \cong \mathcal{O}(E \otimes V)\,.$$

Proof (See Gunning 1967). Parts 1 and 3 are tautological from the construction. For Part 2 we essentially need to show that each point $p \in M$ has a neighbourhood U for which we have an isomorphism

$$f_*\mathcal{O}(L)(U) \cong \underbrace{\mathcal{O}(U) \oplus \cdots \oplus \mathcal{O}(U)}_{m}\,.$$

If p is a regular value of f this is obvious, because $f^{-1}(p)$ consists of m distinct points and since $f' \neq 0$ at all of these, there are m disjoint open sets $U_i \subset \tilde{M}$ on each of which f is a holomorphic diffeomorphism to the same open set $U \subset M$, so here

$$f_*\mathcal{O}(L)(U) = \mathcal{O}(L)(f^{-1}(U)) = \bigoplus_{i=1}^{m} \mathcal{O}(U_i)\,.$$

In the general case, $f^{-1}(p)$ contains branch points, where $f(z)$ looks like $z^k g(z)$. In fact a change of coordinate means we can take neighbourhoods U and $\tilde{U}$ such that the map f is

$$\begin{array}{ccc} \tilde{U} & \to & U \\ z & \mapsto & z^k, \end{array}$$

and a local coordinate on U is given by $w = z^k$. A section of L over $\tilde{U}$ will now look like

$$h(z) = h_0(w) + zh_1(w) + \cdots + z^{k-1}h_{k-1}(w),$$

so the space of sections is the direct sum of k copies of the holomorphic functions on U. The total multiplicity of the branch points still satisfies

$$\sum_{q\in f^{-1}(p)} k_q = \deg f = m$$

so in both cases, local sections of $f_*\mathcal{O}(L)$ look like m local holomorphic functions. It follows that $f_*\mathcal{O}(L)$ is the sheaf of sections of a rank m holomorphic vector bundle E on M. □

Using the above construction, let us work out the basic topological invariant of E, its degree:

Proposition 4.3 *With L and E as above,*

$$\deg E = \deg L + (1 - \tilde{g}) - \deg f(1 - g),$$

where g is the genus of M, and $\tilde{g}$ is the genus of $\tilde{M}$.

Proof (1) For a vector bundle V over M, an argument we have used before shows that

$$\mathrm{H}^0(M, VL_p^{-n}) = 0$$

for large enough n. We simply use the exact sequence of cohomology groups: if $\mathrm{H}^0(M, V)$ and $\mathrm{H}^0(M, VL_p^{-1})$ have the same dimension then all sections of V vanish at p, but a given section can only vanish with some finite multiplicity, so eventually VL_p^{-m} has less sections, and the dimension goes down. Repeating the process, it must finally be zero.

(2) Using Serre duality, it follows that

$$\mathrm{H}^1(M, VL_p^n)^* \cong \mathrm{H}^0(M, V^*KL_p^{-n}) = 0,$$

for large enough n.

(3) Thus we have

$$\mathrm{H}^1(\tilde{M}, Lf^*L_p^n) = 0$$

and

$$\mathrm{H}^1(M, EL_p^n) = 0,$$

for sufficiently large n. Applying the Riemann–Roch theorem to these bundles, and using the above vanishing results, we find

$$\dim \mathrm{H}^0(\tilde{M}, Lf^*L_p^n) = \deg L + mn + (1 - \tilde{g})$$

and

$$\dim \mathrm{H}^0(M, EL_p^n) = \deg E + mn + m(1 - g)\,.$$

By parts 1 and 3 of Proposition 4.2, these dimensions must be the same, which gives us the required result. □

Suppose now that $M = \mathbb{P}^1$, so we are considering a holomorphic map $f : \tilde{M} \to \mathbb{P}^1$. Since $g = 0$, it follows from Proposition 4.3 that $\deg E = 0$ if and only if $\deg L = m + (\tilde{g} - 1)$. If the degree is zero, then E is topologically trivial but may not be holomorphically trivial. The following proposition tells us when the latter is true.

Proposition 4.4 *Suppose* $\deg E = 0$; *then* E *is trivial if and only if*

$$L(-1) = L \otimes f^*\mathcal{O}(-1)$$

has no non-trivial holomorphic sections.

Proof By the corollary to Theorem 4.1, E is trivial if and only if

$$\mathrm{H}^0(\mathbb{P}^1, E(-1)) = 0\,.$$

However, this space of sections is canonically isomorphic to

$$\mathrm{H}^0(\tilde{M}, L(-1))$$

by Proposition 4.2, parts 1 and 3. □

Note that if $\deg L = m + (\tilde{g} - 1)$, then $\deg L(-1) = \tilde{g} - 1$, since the degree of $f^*\mathcal{O}(1)$ is m. So, from Propositions 4.3 and 4.4, if $L(-1)$ is of degree $(\tilde{g} - 1)$ and has no non-zero holomorphic sections, then the holomorphic vector bundle E is trivial. Let us consider this condition more generally.

Recall that on any Riemann surface M of genus g, J^{g-1} is the space of equivalence classes of line bundles of degree $g - 1$ and this is a g-dimensional complex torus, the Jacobian:

$$J^{g-1} \cong \frac{\mathrm{H}^1(M, \mathcal{O})}{\mathrm{H}^1(M, \mathbb{Z})}\,.$$

There is a holomorphic map from $\underbrace{M \times \cdots \times M}_{g-1}$ to J^{g-1} given by

$$(p_1, \ldots, p_{g-1}) \mapsto L_{p_1} \ldots L_{p_{g-1}}\,.$$

The image is called the theta-divisor Θ. If a line bundle L has a non-vanishing section with zeros $p_1, \ldots, p_{g-1}$ then $L \cong L_{p_1} \ldots L_{p_{g-1}}$, so L belongs to the theta-divisor. The image of the $(g-1)$-dimensional product is of codimension one in the g-dimensional Jacobian J^{g-1}.

In the specific situation of $f : \tilde{M} \to \mathbb{P}^1$, it follows that for a generic line bundle $L(-1) \in J^{\tilde{g}-1}$, $f_*\mathcal{O}(L)$ will be the trivial bundle over $\mathbb{P}^1$. If $L(-1) \in \Theta$ then the a_i in the Birkhoff decomposition of E will be non-zero. So in particular, in a continuous family the a_i will jump from zero to non-zero values as one passes through the theta-divisor.

5 Matrix polynomials and Lax pairs

We have seen in the last section that a degree m map

$$f : M \to \mathbb{P}^1$$

and a line bundle $L(-1) \in J^{g-1}\backslash\Theta$ give us a trivial bundle E on $\mathbb{P}^1$. On its own, this is not much information, but we now introduce a new ingredient. We take a section $w \in \mathrm{H}^0(M, f^*\mathcal{O}(n))$.

Let $U \subset \mathbb{P}^1$; then multiplication by w defines a linear map

$$w : \mathrm{H}^0(f^{-1}(U), L) \to \mathrm{H}^0(f^{-1}(U), L(n)).$$

By the definition of E this is a homomorphism

$$W : \mathrm{H}^0(U, E) \to \mathrm{H}^0(U, E(n)).$$

Since W is globally defined, and E is trivial,

$$W : \mathrm{H}^0(\mathbb{P}^1, E) \cong \mathbb{C}^m \to \mathrm{H}^0(\mathbb{P}^1, E(n)) \cong \mathbb{C}^m \otimes \mathrm{H}^0(\mathbb{P}^1, \mathcal{O}(n)).$$

In other words, we have an $m \times m$ matrix-valued holomorphic section of $\mathcal{O}(n)$. We have seen that all sections of $\mathcal{O}(n)$ are polynomials in z of degree $\leq n$, so

$$W = A(z) = A_0 + A_1 z + \cdots + A_n z^n.$$

What we have here is a construction of a matrix polynomial from a line bundle on a Riemann surface. We may ask how to go the other way: what is the role of the Riemann surface M in this picture?

Recall that if p is a regular value of f, then the inverse image $f^{-1}(U)$ of a small neighbourhood U of p in $\mathbb{P}^1$ will consist of m neighbourhoods $U_1, \ldots, U_m$. In particular,

$$\mathrm{H}^0(U, E) = \bigoplus_i \mathrm{H}^0(U_i, L).$$

We can choose a local basis $\{s_1, \ldots, s_m\}$ of sections of E over U which correspond to sections $\{\sigma_1, \ldots, \sigma_m\}$ of L over $f^{-1}(U)$ which satisfy $\sigma_i|_{U_j} = 0$ for $i \neq j$. Since E is trivial, the s_i are local vector-valued functions. But w acts on the sections σ_i by scalar multiplication, and so

$$A(z)s_i = w|_{U_i} s_i,$$

thus the m locally defined functions $w|_{U_i} \circ f^{-1}|_U$ are eigenvalues of $A(z)$. In particular, by continuity, at every point of M, the value of w satisfies the characteristic equation $\det(w - A(z)) = 0$. This is an algebraic relation between sections of line bundles on M and describes M as an algebraic curve as follows.

The section w of $f^*\mathcal{O}(n)$ defines the following commutative diagram

$$\begin{array}{ccc} & & \mathcal{O}(n) \\ & {\scriptstyle w}\nearrow & \downarrow{\scriptstyle \pi} \\ M & \xrightarrow[f]{} & \mathbb{P}^1 \end{array}$$

We denote by S the total space of the line bundle $\mathcal{O}(n)$: S is a two-dimensional complex manifold. Then the image of M in S is given by the equation

$$\det(w - A(z)) = w^m + a_1(z)w^{m-1} + \cdots + a_m(z) = 0$$

and is called the *spectral curve* of $A(z)$.

Let us assume for simplicity now that $w(M)$ embeds M as a one-dimensional submanifold of S.

If $M \subset S$ determines the spectrum of $A(z)$, what, we may ask, has the line bundle which enters into the construction got to do with $A(z)$?

The matrix $A(z)$ acts on $\mathrm{H}^0(M, L) = \mathbb{C}^m$ and evaluating a section of L at a point gives a surjective homomorphism of vector bundles

$$M \times \mathbb{C}^m \to L$$

which is preserved by w. Taking duals, $L^* \subset M \times (\mathbb{C}^m)^*$ is the eigenspace of $A^t(z)$ corresponding to the eigenvalue w.

Remark The point about this construction is that the eigenvalues of the polynomial matrix $A(z)$ are not single-valued functions of z: for each z there are m solutions to $\det(w - A(z)) = 0$. This means there is an m-fold covering of $\mathbb{P}^1$ on which we do have a single-valued eigenvalue w. This covering is the spectral curve S, and if $A(z)$ is generic the eigenspace is just one-dimensional and defines the line bundle L. In Galois theory terms this is like producing a field extension in which the algebraic equation $\det(w - A(z)) = 0$ has a root: in fact the field of meromorphic functions on the spectral curve is a finite extension of the field of rational functions in z which fulfils precisely that role.

We thus have the following correspondence:

- the spectrum of $A(z)$ is a Riemann surface M
- the eigenspace of $A^t(z)$ is a line bundle on M

and the direct image construction relates one viewpoint to the other. We can put this more formally as a proposition:

Proposition 5.1 *Let X be the space of all $m \times m$ matrix polynomials $A(z)$ with smooth spectral curve Σ. Then $PGL(m, \mathbf{C})$ acts freely on X by conjugation and the quotient can be identified with $J^{g-1}(\Sigma)\backslash\Theta$.*

We are now in a position to discuss the origin of Lax pair equations and the role of Riemann surfaces and line bundles in their solution. Recall that we consider equations of the form

$$\frac{\mathrm{d}A}{\mathrm{d}t} = [A, B]$$

where $A(z)$ and $B(z)$ are matrix polynomials. It is easy to see that the spectrum of $A(z)$ is preserved if A satisfies an equation of this form. From our point of view it means that as t varies, the eigenspace bundle varies in a complex torus. We shall now show that if a line bundle follows a straight line motion in the torus, there is a basis for which $A(z)$ evolves as a Lax pair equation, where $B(z)$ has a specific form.

We begin then with $f : M \to \mathbb{P}^1$ and $L_t(-1) \in J^{g-1}\backslash\Theta$ varying linearly in t. To go further, we need a good description of line bundles on M. Recall that M sits inside the total space S of the line bundle $\mathcal{O}(n)$ over $\mathbb{P}^1$ given by the transition function z^n. We can use the standard coordinate patches on S, and then restrict to M. In particular, we can write $S = U \cup \tilde{U}$, where U and $\tilde{U}$ are both isomorphic to $\mathbb{C}^2$ with coordinates (z, w) and $(\tilde{z}, \tilde{w})$ respectively. On the overlap $\mathbb{C}^* \times \mathbb{C}$, these coordinates will be related by

$$\begin{aligned} \tilde{z} &= z^{-1} \\ \tilde{w} &= wz^{-n}. \end{aligned}$$

Furthermore, U and $\tilde{U}$ are good open sets in the sense that we can use this covering to calculate cohomology of sheaves of sections of line bundles on S.

Recall that $w(M) \subset S$ is given by

$$\det(w - A(z)) = w^m + a_1(z)w^{m-1} + \cdots + a_m(z) = 0,$$

thus it is the zero set of a holomorphic section of $\pi^*\mathcal{O}(mn)$ on S (where π is the projection $\pi : S \to \mathbb{P}^1$). Therefore there is a short exact sequence

$$0 \to \mathcal{O}_S(-mn) \overset{\det(w-A(z))}{\longrightarrow} \mathcal{O}_S \to \mathcal{O}_M \to 0\,.$$

Taking the corresponding long exact sequence we get

$$\mathrm{H}^1(S, \mathcal{O}(-mn)) \overset{\det(w-A(z))}{\to} \mathrm{H}^1(S, \mathcal{O}) \to \mathrm{H}^1(M, \mathcal{O}) \to \mathrm{H}^2(S, \mathcal{O}(-mn))\,.$$

The last of these groups vanishes, because we covered S with only two open sets, so there are no three-fold intersections. This means that $\mathrm{H}^1(M, \mathcal{O})$ can be described as $\mathrm{H}^1(S, \mathcal{O})$ modulo the image of $\det(w - A(z))$. Now $\mathrm{H}^1(S, \mathcal{O})$ is given by holomorphic functions on $U \cap \tilde{U} = \mathbb{C}^* \times \mathbb{C}$ modulo functions which extend to U and $\tilde{U}$. So overall, $\mathrm{H}^1(M, \mathcal{O})$ can be described as functions

$$\sum_{l=-\infty}^{\infty} \sum_{k=0}^{\infty} a_{kl} w^k z^l$$

on $\mathbb{C}^* \times \mathbb{C}$ modulo functions

$$\sum_{l=0}^{\infty} \sum_{k=0}^{\infty} a_{kl} w^k z^l$$

on U, functions

$$\sum_{l=0}^{\infty} \sum_{k=0}^{\infty} \tilde{a}_{kl} \tilde{w}^k \tilde{z}^l = \sum_{l=0}^{\infty} \sum_{k=0}^{\infty} \tilde{a}_{kl} (wz^{-n})^k (z^{-1})^l = \sum_{l=0}^{\infty} \sum_{k=0}^{\infty} \tilde{a}_{kl} w^k z^{-nk-l}$$

on $\tilde{U}$, and modulo the equation

$$\det(w - A(z)) = w^m + a_1(z)w^{m-1} + \cdots + a_m(z) = 0\,.$$

Thus we find that we can represent any class in $\mathrm{H}^1(M, \mathcal{O})$ by

$$\beta(w, z) = \sum_{i=1}^{m-1} \frac{b_i(z) w^i}{z^N},$$

where N is some positive integer and the $b_i(z)$'s are polynomials.

Now $\exp \beta(w,z)$ will give us a class in $\mathrm{H}^1(M,\mathcal{O}^*)$ corresponding to a degree zero line bundle. Therefore using $M \cap U$ and $M \cap \tilde{U}$ as an open cover for M, we find we can express any line bundle L of degree d as a bundle with transition function $\exp \beta(w,z)c$, where c is the transition function for some fixed line bundle of degree d. Then to make a linear variation in L we merely need to vary our element of $\mathrm{H}^1(M,\mathcal{O})$ linearly. Thus we get a family of line bundles L_t corresponding to the transition functions $\exp(t\beta)c$ where c is independent of t.

Suppose we have a family of sections varying with t, each section in

$$\mathrm{H}^0(M, L_t) = \mathrm{H}^0(\mathbb{P}^1, E_t) = \mathbb{C}^m.$$

These will be given by functions $s(t)$ and $\tilde{s}(t)$ on U and $\tilde{U}$ respectively, which are related on the overlap $U \cap \tilde{U}$ by

$$s(t) = \exp(t\beta(w,z))c\tilde{s}(t).$$

So take a holomorphically varying basis of global sections

$$\sigma_1(t), \dots, \sigma_m(t)$$

of L_t, then this is represented by functions $s_i, \tilde{s}_i$ for $1 \le i \le m$, with

$$s_i(t) = \exp(t\beta(w,z))c\tilde{s}_i(t).$$

Differentiating this equation with respect to t we get

$$\frac{\partial s_i}{\partial t} = \beta s_i + \exp(t\beta)c\frac{\partial \tilde{s}_i}{\partial t}\,.$$

Now by the definition of $A(z)$,

$$ws_i = \sum_j A_{ji}s_j\,,$$

where A_{ij} is the matrix of $A(z)$ in this basis. Since $\beta(w,z)$ is polynomial in w, we can then write the above equation as

$$\frac{\partial s_i}{\partial t} = \sum_j \beta(A,z)_{ji}s_j + \exp(t\beta)c\frac{\partial \tilde{s}_i}{\partial t}\,.$$

Now split the finite matrix Laurent series $\beta(A(z), z)$ into polynomials β^+ in z and β^- in z^{-1} (put the constant term in β^+):

$$\beta(A(z), z) = \beta^+ + \beta^-.$$

We then have

$$\begin{aligned}
\frac{\partial s_i}{\partial t} - \sum_j \beta^+_{ji} s_j &= \sum_j \beta^-_{ji} s_j + \exp(t\beta) c \frac{\partial \tilde{s}_i}{\partial t} \\
&= \exp(t\beta) c \left(\frac{\partial \tilde{s}_i}{\partial t} - \sum_j \beta^-_{ji} \tilde{s}_j \right)
\end{aligned}$$

and so since one side is holomorphic in z and the other in z^{-1}, this defines a global section $u_i(t)$ of L_t. But $\sigma_1, \ldots, \sigma_m$ were chosen to be a basis, so there is a matrix $C(t)$ depending holomorphically on t alone such that

$$u_i = \sum_j C_{ji} \sigma_j$$

and we have

$$\frac{\partial s_i}{\partial t} - \sum_j \beta^+_{ji} s_j = \sum_j C_{ji} s_j \,.$$

Recall the relation

$$w s_i = \sum_j A_{ji}(z) s_j \,.$$

Differentiating this, and noting that the spectrum does not vary with t, we get

$$\begin{aligned}
w \frac{\partial s_i}{\partial t} &= \sum_j \left(\frac{\partial A}{\partial t} \right)_{ji} s_j + \sum_j A_{ji} \frac{\partial s_j}{\partial t} \\
&= \sum_j \left(\frac{\partial A}{\partial t} \right)_{ji} s_j + \sum_{j,k} A_{ji} (\beta^+_{kj} + C_{kj}) s_k \,.
\end{aligned}$$

On the other hand, the left-hand side can be written

$$w \sum_j (\beta^+_{ji} + C_{ji}) s_j = \sum_{j,k} (\beta^+_{ji} + C_{ji}) A_{kj} s_k \,,$$

and equating these two expressions we find

$$\frac{\partial A}{\partial t} = [A, \beta^+ + C(t)]$$

This is in general a time-dependent equation, but we derived it by making an arbitrary choice of moving basis for $H^0(M, L_t)$. If $\tilde{A} = P^{-1}AP$, then

$$\frac{\mathrm{d}\tilde{A}}{\mathrm{d}t} = [\tilde{A}, \beta^+(\tilde{A}, z)]$$

if

$$\frac{\mathrm{d}P}{\mathrm{d}t} = -CP\,.$$

Thus solving this linear equation gives a natural moving frame in which the equation is of autonomous Lax form

$$\frac{\mathrm{d}A}{\mathrm{d}t} = [A, B]\,.$$

We conclude that if L_t varies linearly in t, then there exists a basis for the trivial bundle on $\mathbb{P}^1$ such that $A(z)$ satisfies a Lax pair equation

$$\frac{\mathrm{d}A}{\mathrm{d}t} = [A, B]\,.$$

This solution becomes singular when the linear flow hits the theta-divisor in J^{g-1}.

The procedure we have adopted here gives a concrete way of interpreting linear flows on the space of line bundles as solutions to Lax pair equations. We represent the direction of the flow by a cohomology class

$$\beta(z, w) = \sum_{i=1}^{m-1} \frac{b_i(z)w^i}{z^N},$$

then take B as the polynomial part

$$B = \beta(A(z), z)^+\,.$$

Note that B is not unique: that is in the nature of cohomology, which is a space of equivalence classes. In particular, we can add onto B any multiple of a power of A. Let us look at a couple of examples.

Examples

(1) We take a dimensional reduction of the self-dual Yang–Mills equations. The full equations are

$$\begin{aligned}[\nabla_0, \nabla_1] &= [\nabla_2, \nabla_3] \\ [\nabla_0, \nabla_2] &= [\nabla_3, \nabla_1] \\ [\nabla_0, \nabla_3] &= [\nabla_1, \nabla_2],\end{aligned}$$

where

$$\nabla_i = \frac{\partial}{\partial x_i} + A_i$$

are covariant derivatives corresponding to a connection A_i in coordinates (x_0, x_1, x_2, x_3). We shall look for solutions depending only on $t = x_0$, i.e. $A_i = T_i(t)$. These are solutions invariant under translation in the x_1, x_2, x_3-directions. We can make a gauge transformation such that ∇_0 becomes simply $\mathrm{d}/\mathrm{d}t$. Then the SDYM equations become

$$\begin{aligned}\frac{\mathrm{d}T_1}{\mathrm{d}t} &= [T_2, T_3] \\ \frac{\mathrm{d}T_2}{\mathrm{d}t} &= [T_3, T_1] \\ \frac{\mathrm{d}T_3}{\mathrm{d}t} &= [T_1, T_2],\end{aligned}$$

which are known as Nahm's equations. Let

$$A(z) = (T_1 + \mathrm{i}T_2) + 2T_3 z - (T_1 - \mathrm{i}T_2)z^2 .$$

Then

$$\begin{aligned}\frac{\mathrm{d}A}{\mathrm{d}t} &= [T_2 - \mathrm{i}T_1, T_3] + 2[T_1, T_2]z - [T_2 + \mathrm{i}T_1, T_3]z^2 \\ &= [A, -\mathrm{i}T_3 + \mathrm{i}(T_1 - \mathrm{i}T_2)z] \\ &= \left[A, -\mathrm{i}\left(\frac{A}{z}\right)^+\right].\end{aligned}$$

We recognize here the form we have discussed above. The matrix $A(z)$ is quadratic in z so $n = 2$ and the direction of the linear flow is given by

$$\beta(z, w) = -\mathrm{i}\,\frac{w}{z}.$$

A special case is when the T_i are of the form

$$T_1 = u_1 \begin{pmatrix} \mathrm{i} & 0 \\ 0 & -\mathrm{i} \end{pmatrix}, \quad T_2 = u_2 \begin{pmatrix} 0 & \mathrm{i} \\ \mathrm{i} & 0 \end{pmatrix}, \quad T_3 = u_3 \begin{pmatrix} 0 & -1 \\ 1 & 0 \end{pmatrix}.$$

Then Nahm's equations become

$$\begin{aligned} \dot{u}_1 &= 2u_2u_3 \\ \dot{u}_2 &= 2u_3u_1 \\ \dot{u}_3 &= 2u_1u_2 . \end{aligned}$$

These are equivalent to the equations of a rigid body pivoted about its centre of mass. In this example, the Riemann surface M has genus 1, and our construction is compatible with the classical solution of these equations using elliptic functions.

(2) A well-known classical integrable system is the equation for geodesics on an ellipsoid

$$\sum_i \beta_i x_i^2 = 1 .$$

This can be approached as follows (Adler and van Moerbeke 1980): let β be the matrix with diagonal entries $\beta_1, \ldots, \beta_n$. Then if $A(z)$ is of the form

$$A(z) = -x \otimes x + z(x \otimes y - y \otimes x) + z^2\beta^{-1} ,$$

the equations for the geodesic flow can be put in Lax form with

$$B = (z^3 A^{-1})^+$$

using the Cayley–Hamilton theorem to express A^{-1} as a polynomial in A.

6 Completely integrable Hamiltonian systems

The geometrical way of solving Lax pair equations outlined above naturally presents us with two sets of variables: the coefficients of the equation of the spectral curve, which are conserved quantities for the flow, and the coordinates of a complex torus, on which the flow is linear. This structure can be seen as part of the more traditional idea of integrable systems, based on the integrability of mechanical systems using Hamiltonian methods. This point of view also leads us to uncharted territory: integrable systems where it becomes very difficult to write down the equations explicitly even though the geometry is quite clear. We begin with a resumé of symplectic geometry.

A symplectic manifold X^{2n} is a manifold with a non-degenerate closed 2-form ω. We can adopt this definition for complex manifolds if we assume that ω is holomorphic. This skew-form gives us an isomorphism between the tangent bundle and cotangent bundle of X, so that given a function $H : X \to \mathbb{C}$ (known as a Hamiltonian) we obtain a vector field X_H (known as a Hamiltonian vector field) characterized by $\omega(X_H, Y) = Y(H)$ for a vector field Y. Given two functions f and g on X, their Poisson bracket is defined as

$$\{f, g\} := X_f(g) = -X_g(f),$$

and

$$X_{\{f,g\}} = [X_f, X_g].$$

Apart from the cotangent bundle of any manifold, the most natural example of a symplectic manifold is any coadjoint orbit of a Lie group G. If $\xi \in \mathfrak{g}^*$, the dual of the Lie algebra of G, then the symplectic structure at ξ is given by

$$\omega_\xi(X, Y) = \xi([X, Y]),$$

where X and $Y \in \mathfrak{g}$ define tangent vectors to the orbit at ξ. This is a symplectic manifold with an action of a Lie group preserving the symplectic structure. The fact that it sits inside $\mathfrak{g}^*$ is part of a general property of symplectic manifolds X with group actions: the momentum map is an equivariant map

$$\mu : M \to \mathfrak{g}^*$$

and in general $\mu^{-1}(0)/G$ is another symplectic manifold, the symplectic quotient.

We begin now with $G = \mathrm{SL}_m(\mathbb{C})$ and identify the Lie algebra with its dual using the bilinear form $\mathrm{tr}(AB)$. Take a product of orbits

$$O_1 \times \cdots \times O_k$$

of elements of $\mathfrak{g}$ with distinct eigenvalues. The dimensions of each of these orbits is

$$\dim O_i = (m^2 - 1) - (m - 1) = m(m - 1)$$

since $\dim \mathrm{SL}_m(\mathbb{C}) = m^2 - 1$ and the diagonal matrices with trace zero have dimension $m-1$. We now perform a symplectic reduction on this product. The moment map for the action of G is given by

$$\begin{array}{lcl} O_1 \times \cdots \times O_k & \to & \mathfrak{g} \\ (R_1, \ldots, R_k) & \mapsto & \sum_i R_i \,. \end{array}$$

The symplectic quotient is then

$$X = \{(R_1, \ldots, R_k) : \sum_i R_i = 0\}/\mathrm{SL}_m(\mathbb{C}),$$

and this is a symplectic manifold of dimension

$$2N = k(m^2 - m) - 2(m^2 - 1)\,. \tag{6.1}$$

We are going to relate this symplectic picture to matrix polynomials $A(z)$. Let $p(z) = (z - \alpha_1) \ldots (z - \alpha_k)$ and

$$A(z) = p(z) \sum_{i=1}^{k} \frac{R_i}{(z - \alpha_i)}\,.$$

Then $A(z)$ is a matrix-valued polynomial of degree $(k - 2)$, as the leading term $\sum_i R_i$ vanishes. Now

$$\det(w - A(z)) = w^m + a_2 w^{m-2} + a_3 w^{m-3} + \cdots + a_m = P(z, w)$$

(note that $a_1 = \mathrm{tr} A = 0$, since we are dealing with matrices in $\mathfrak{sl}_m(\mathbb{C})$, i.e. matrices that have zero trace).

Each a_i is a polynomial in z, but not arbitrary, since $R_i \in O_i$. Thus the eigenvalues λ_i of $A(\alpha_i)$ are fixed in advance and $P(\alpha_i, \lambda_i^{(j)}) = 0$. Subject to these constraints, the coefficients are independent functions on X. If we fix them to be a constant (i.e. fix the spectral curve of $A(z)$) then we know from Proposition 5.1 that the subspace of X so defined can be identified with $J^{g-1} \backslash \Theta$, so Jacobians of Riemann surfaces are appearing inside this symplectic manifold. We shall see that the Hamiltonian vector fields of the functions on X which are coefficients of the characteristic polynomial actually give us Lax pair equations, linearized on these Jacobians. We shall only do the calculation for the simplest case, but the picture holds in general (Adams, Harnad, and Hurtubise 1993).

Consider

$$R(z) = \frac{1}{p(z)}A(z) = \sum_{i=1}^{k} \frac{R_i}{(z-\alpha_i)},$$

and the function H defined by the residue at $z = \alpha_k$ of $\operatorname{tr}(R(z))^2$, i.e.

$$H = \sum_{j\neq k} \frac{\operatorname{tr}(R_j R_k)}{(\alpha_k - \alpha_j)}.$$

Now take a tangent vector Y on the product of orbits, representing a tangent in the quotient. This is defined by $\dot{R}_i = [Y_i, R_i]$ for $Y_i \in \mathfrak{g}$. Next differentiate H in the direction of Y. We obtain

$$Y(H) = \sum_{j\neq k} \frac{\operatorname{tr}([Y_j, R_j]R_k)}{(\alpha_k - \alpha_j)} + \sum_{j\neq k} \frac{\operatorname{tr}(R_j[Y_k, R_k])}{(\alpha_k - \alpha_j)}.$$

If X_H is the Hamiltonian vector field of H, it is defined by $X_i \in \mathfrak{g}$ with $\dot{R}_i = [X_i, R_i]$, and by the definition of the symplectic form on products of coadjoint orbits,

$$Y(H) = \omega(X_H, Y) = \sum_j \operatorname{tr}(R_j[X_j, Y_j]) = \sum_j \operatorname{tr}([Y_j, R_j]X_j).$$

By comparing with the formula for $Y(H)$ above, we see that

$$X_j = \frac{R_k}{\alpha_k - \alpha_j} \quad (j \neq k), \qquad X_k = \sum_{j\neq k} \frac{R_j}{\alpha_k - \alpha_j}$$

so that the Hamiltonian flow gives the differential equation

$$\begin{aligned} \frac{\mathrm{d}R}{\mathrm{d}t} &= \sum_{j\neq k} \frac{[R_k, R_j]}{(z-\alpha_j)(\alpha_k - \alpha_j)} + \sum_{j\neq k} \frac{[R_j, R_k]}{(z-\alpha_k)(\alpha_k - \alpha_j)} \\ &= \sum_{j\neq k} \frac{[R_j, R_k]}{(z-\alpha_k)(z-\alpha_j)} \\ &= \left[R, \frac{R_k}{z-\alpha_k}\right] \end{aligned}$$

and this is in Lax form. To see that it is linear on the Jacobian, choose the coordinate z such that $\alpha_k = 0$. Then the Lax equation is

$$\frac{\mathrm{d}A}{\mathrm{d}t} = \left[A, \frac{R_k}{z}\right] = \left[A, \left(\frac{A}{p'(0)z}\right)^{-}\right] = -\left[A, \left(\frac{A}{p'(0)z}\right)^{+}\right].$$

As in the previous section, we see that the Hamiltonian flow of the function H is tangential to the Jacobians in X and is moreover linear. In other words the vector field has constant coefficients with respect to the flat coordinates on the torus. We have k such functions $H_1, \ldots, H_k$ and since they correspond to constant vector fields on the torus, they commute, or equivalently the functions H_i, H_j on X Poisson-commute: $\{H_i, H_j\} = 0$. If we perform the calculations above for the coefficients of $\mathrm{tr}(R(z)^p)$ for all p then one can show similarly that the corresponding functions Poisson-commute. This brings us face to face with the notion of complete integrability in a Hamiltonian setting.

Definition 6.1 *A completely integrable Hamiltonian system is a symplectic manifold X^{2N} with N Hamiltonian functions $H_1, \ldots, H_N$ such that $\{H_i, H_j\} = 0$ and $dH_1 \wedge \ldots \wedge dH_N \not\equiv 0$.*

We shall have shown that our example is a (complex) completely integrable Hamiltonian system if the number of independent coefficients of the characteristic polynomial

$$\det(w - A(z)) = w^m + a_2 w^{m-2} + a_3 w^{m-3} + \cdots + a_m = P(z, w)$$

is precisely N. In fact a numerical coincidence shows this to be true. Note that each $a_i(z)$ is of degree $i(k-2)$ since $A(z)$ is of degree $(k-2)$. If we chose them arbitrarily, the number of degrees of freedom would be

$$\sum_{i=2}^{m} (i(k-2)+1) = \left(\frac{m(m+1)}{2} - 1\right)(k-2) + (m-1)\,.$$

But we have the constraints

$$P(\alpha_i, \lambda_i^{(j)}) = 0$$

and so the actual number of degrees of freedom is

$$\left(\frac{m(m+1)}{2} - 1\right)(k-2) + (m-1) - (m-1)k = k\frac{m(m-1)}{2} - (m^2 - 1)\,.$$

But this is indeed precisely *half* the dimension of the symplectic reduction (see eqn 6.1)!

This result is part of a bigger picture, which involves the study of the moduli space of *filtered Higgs bundles* (Hitchin 1987; Simpson 1990, 1992). In the above situation we should write more invariantly

$$R(z) = \sum_j \frac{R_j}{z - \alpha_j} \, \mathrm{d}z$$

so that the R_j are the residues of a meromorphic matrix-valued differential on $\mathbb{P}^1$. Moreover the fact that the sum of the residues is zero is the vanishing of the moment map for the symplectic reduction. The generalization consists of replacing $R(z)$, which is a meromorphic form on $\mathbb{P}^1$ with values in the endomorphisms of the trivial bundle, with a 1-form on an arbitrary Riemann surface M with values in $\operatorname{End} E$ for an arbitrary holomorphic vector bundle E. The numerical miracle is repeated and we get many completely integrable systems this way. To illustrate it, let us go to the opposite extreme from the above example: we shall use an arbitrary compact Riemann surface M of genus $g > 1$ and a *holomorphic* form with values in $\operatorname{End} E$—no poles at all.

We need to consider the space of equivalence classes of holomorphic rank m vector bundles on M. This is not such a good object unless we make the supposition that the bundle is what is called stable. This is a generic condition and one of its consequences is that the only global endomorphisms of the bundle are the scalars. There is, for stable bundles, a good space of equivalence classes, which is a complex manifold $\mathcal{M}^n$.

The dimension of $\mathcal{M}^n$ can be calculated by looking at infinitesimal deformations of the holomorphic structure. If $g_{\alpha\beta}$ is a set of transition matrices in a 1-parameter family of holomorphic bundles then

$$g_{\alpha\beta}^{-1}\dot{g}_{\alpha\beta}$$

defines a class in the sheaf cohomology group $\mathrm{H}^1(M, \operatorname{End} E)$. This space is then naturally identified with the tangent space to $\mathcal{M}^n$ at $[E]$. We can calculate its dimension from the Riemann–Roch theorem:

$$\begin{aligned} &\dim \mathrm{H}^0(M, \operatorname{End} E) - \dim \mathrm{H}^1(M, \operatorname{End} E) \\ &= \deg \operatorname{End} E + \operatorname{rk} \operatorname{End} E (1 - g)\,. \end{aligned}$$

By stability $\dim \mathrm{H}^0(M, \mathrm{End}\, E) = 1$, just the constant scalars, and since $\mathrm{End}\, E$ has a nondegenerate holomorphic bilinear form on it $\deg \mathrm{End}\, E = 0$. Thus

$$\dim \mathcal{M} = \dim \mathrm{H}^1(M, \mathrm{End}\, E) = 1 - m^2(1-g)\,. \qquad (6.2)$$

Now $\mathcal{M}$ itself is not a symplectic manifold, but we take its cotangent bundle $T^*\mathcal{M}$ with its canonical symplectic structure. Since the tangent space to $\mathcal{M}$ at $[E]$ is $\mathrm{H}^1(M, \mathrm{End}\, E)$, Serre duality gives us the cotangent space as

$$\mathrm{H}^0(M, \mathrm{End}\, E \otimes K)\,.$$

A point in the cotangent bundle is thus (up to equivalence) a vector bundle E and a holomorphic section A of $\mathrm{End}\, E \otimes K$. Locally, A is just an $m \times m$ matrix with values in the canonical bundle, and we can take its characteristic polynomial

$$\det(w - A) = w^m + a_1 w^{m-1} + \cdots + a_m$$

where now instead of each coefficient a_i being a polynomial, it is a global holomorphic section of the line bundle K^i. We need to calculate the dimension of $\mathrm{H}^0(M, K^i)$.

Now when $i = 1$, by definition $\dim \mathrm{H}^0(M, K) = g$ and by Serre duality $\dim \mathrm{H}^1(M, K) = 1$. So by Riemann–Roch,

$$\deg K = 2g - 2\,.$$

If $g > 1$ this is positive. Now by Serre duality,

$$\mathrm{H}^1(M, K^i)^* \cong \mathrm{H}^0(M, K^{1-i})$$

so for $i > 1$, the line bundle K^{1-i} has negative degree and no sections. Riemann–Roch for the line bundle K^i then gives

$$\dim \mathrm{H}^0(M, K^i) = \deg K^i + (1-g) = (2i-1)(g-1)\,.$$

The number of degrees of freedom in choosing the characteristic polynomial of A is thus

$$g + \sum_{i=2}^{m} (2i-1)(g-1) = (m^2-1)(g-1) + g = 1 - m^2(1-g)$$

and this is exactly the dimension of $\mathcal{M}$ or *half* the dimension of the symplectic manifold $T^*\mathcal{M}$ (see eqn 6.2).

In this case, the equation $\det(w - A) = 0$ defines a spectral curve $\tilde{M}$ in the total space of the line bundle K of M, and A is determined by choosing a line bundle on $\tilde{M}$ and making the direct image construction of Section 4. Here again the fibre is an open set in a complex torus and the Hamiltonian flows are linear.

In this general framework, we see integrable systems associated to Riemann surfaces but it seems impossible to attempt to write them down explicitly. Only the geometry remains.

Bibliography

Adams, M. R., Harnad, J., and Hurtubise, J. (1993). Darboux coordinates and Liouville–Arnold integration in loop algebras. *Comm. Math. Phys.*, **155**, 385–413.

Adler, M. and van Moerbeke, P. (1980). Completely integrable systems, Euclidean Lie algebras and curves; and: Linearization of Hamiltonian systems, Jacobi varieties and representation theory. *Adv. Math.*, **38**, 267–379.

Audin, M. (1996). *Spinning tops.* Cambridge University Press.

Farkas, H. M. and Kra, I. (1980). *Riemann surfaces.* Springer-Verlag. New York.

Griffiths, P. and Harris, J. (1978). *Principles of algebraic geometry.* Wiley, New York.

Gunning, R. C. (1966). *Lectures on Riemann surfaces.* Princeton University Press.

Gunning, R. C. (1967). *Lectures on vector bundles over Riemann surfaces.* Princeton University Press.

Gunning, R. C. (1972). *Lectures on Riemann surfaces, Jacobi varieties*, Princeton University Press.

Hitchin, N. J. (1987). Stable bundles and integrable systems. *Duke Math. J.*, **54**, 91–114.

Simpson, C. (1990). Harmonic bundles on noncompact curves. *J. Amer. Math. Soc.*, **3**, 713–70.

Simpson, C. (1992). Higgs bundles and local systems. *Inst. Hautes Etudes Sci. Publ. Math.*, **75**, 5–95.

3

Integrable systems and inverse scattering

Graeme Segal

1 Solitons and the KdV equation

The original soliton, or 'solitary wave', was the bow-wave of a canal-barge observed in 1834 near Edinburgh by the marine engineer John Scott Russell. The barge was stopped suddenly, and Scott Russell, galloping alongside on his horse, saw the bow-wave travel for miles along the canal without changing its shape, size, or speed.[1]

Only in 1895, Korteweg and de Vries found the equation

$$-4\frac{\partial u}{\partial t} = \frac{\partial^3 u}{\partial x^3} + 6u\frac{\partial u}{\partial x} \tag{1.1}$$

describing waves in a shallow canal. One does not expect a non-linear equation like this to describe a non-dispersing wave. For if one neglects the non-linear term and looks for a solution of

$$\frac{\partial u}{\partial t} = \frac{\partial^3 u}{\partial x^3}$$

of the form $\cos(kx - \omega t)$ one finds that $\omega = k^3$, so that Fourier components of different frequency ω travel with different speeds $\omega/k = \omega^{2/3}$, and so spread out. Taking account of the non-linearity ought to make the behaviour more complicated. Nevertheless, somehow the non-linearity exactly compensates for the dispersion, and the KdV equation (1.1) has the solution

$$u(x,t) = 2k^2 \mathrm{sech}^2(kx - k^3 t),$$

describing a lump moving with a constant speed k^2 which is proportional to its amplitude.

---

[1]For a history of the subject, see Newell (1983).

There matters stood until the 1960s, when it was discovered by numerical integration that if one starts with an initial configuration $u(x) = u(x, 0)$ which is close to a sum

$$\sum 2p_i^2 \operatorname{sech}^2(p_i(x - a_i))$$

of solitons of different sizes centred at widely separated points a_i of the line, then after a long time t the solution $u(x, t)$ is very nearly of the same form, as if the individual solitons had sailed through each other without interacting, except that when a bigger one overtakes a smaller one it suffers a certain delay. Moreover, if we start with any initial function u which decays rapidly at infinity, and let it evolve for a long time, then what we obtain looks roughly like a collection of widely separated solitons.

Once again, this behaviour is unexpected, as one cannot usually superpose solutions of a non-linear equation. The survival of the individual lumps suggests that the equation must lead to a large number of conservation laws, and that is indeed the case. In 1967 the whole picture was beautifully explained by reformulating the KdV equation as a *Lax equation*, and using the methods of 'inverse-scattering' theory. I shall first say a little about inverse-scattering, and then about its application to the KdV equation.

Inverse scattering

Waves on a string are described by the wave equation $\ddot{\varphi} - \varphi'' = 0$. (Here $\dot{\varphi} = \partial\varphi/\partial t$, and $\varphi' = \partial\varphi/\partial x$.) If we constrain the string in a certain region $-A \leqslant x \leqslant A$ the equation of motion will become

$$\ddot{\varphi} - \varphi'' + u\varphi = 0,$$

where $u : \mathbb{R} \to \mathbb{R}$ is a smooth function with support in $[-A, A]$ which describes the constraint or 'obstacle'.

Let us consider a wave-packet travelling towards the obstacle, i.e. suppose that for $t \ll 0$ we have

$$\varphi(x, t) = f(x - t),$$

where f has compact support. For a genuine obstacle, with $u \geqslant 0$ everywhere, the solution for $t \gg 0$ will be the sum of a 'reflected' and a 'transmitted' wave:

$$\varphi(x, t) = f_{\mathrm{R}}(x + t) + f_{\mathrm{T}}(x - t), \tag{1.2}$$

where f_R and f_T have compact support. 'Inverse scattering theory' is the art of reconstructing the obstacle from the reflected waves, i.e. of determining the function u from the map $f \mapsto f_R$. It becomes a little more complicated if we do not have $u \geqslant 0$ everywhere, for then the obstacle can 'capture' part of the incoming wave, and (1.2) must be replaced by

$$\varphi(x,t) = f_R(x+t) + f_T(x-t) + \sum \psi_k(x)\cosh(\omega_k t + \delta_k)$$

for $t \gg 0$, where the ψ_k are solutions of the equation

$$-\psi'' + u\psi = -\omega_k^2\psi$$

which decay rapidly as $|x| \to \infty$. Such solutions, called 'bound states', exist for only finitely many values of ω_k. When bound states exist one needs more information than just the map $f \mapsto f_R$ in order to reconstruct the function u, but not very much more, in fact only the numbers ω_k and, for each k, a number ρ_k such that

$$\psi_k(x) = \begin{cases} e^{\omega_k x} & \text{for } x \ll 0 \\ \rho_k e^{-\omega_k x} & \text{for } x \gg 0\,. \end{cases}$$

It is possible, however, to have 'reflectionless potentials' u for which f_R is zero for all f. These turn out to be the 'pure multi-solitons' of the KdV equation, parametrized by (ω_k, ρ_k).

In practice, scattering is treated in a time-independent way. We are really studying the operator

$$L_u = -\left(\frac{d}{dx}\right)^2 + u \tag{1.3}$$

which acts on functions on the line. We look for eigenfunctions, i.e. solutions of

$$L_u\varphi_\lambda = \lambda^2\varphi_\lambda$$

(for $\lambda \in \mathbb{R}$), which are of the form

$$\varphi_\lambda(x) = \begin{cases} e^{i\lambda x} + R_\lambda e^{-i\lambda x} & \text{for } x \ll 0, \\ T_\lambda e^{i\lambda x} & \text{for } x \gg 0\,. \end{cases}$$

The functions R_λ and T_λ are called the *reflection* and *transmission* coefficients respectively. It turns out that $\lambda \mapsto R_\lambda$ is a smooth function of rapid decrease on the line, and it determines the map $f \mapsto f_{\mathrm{R}}$. (See Section 5.) It is a kind of non-linear Fourier transform of u: in fact for small u we have (cf. Proposition 6.5)

$$R_\lambda = \int_{\mathbb{R}} u(x)\mathrm{e}^{-\mathrm{i}\lambda x}\,\mathrm{d}x + O(u^2)\,.$$

Example If u is the 'soliton' $-2k^2 \operatorname{sech}^2 kx$ already mentioned, then

$$\varphi_\lambda(x) = \mathrm{e}^{\mathrm{i}\lambda x}\left\{1 + \frac{\mathrm{i}k}{\lambda - \mathrm{i}k}(1 + \tanh kx)\right\}.$$

For this eigenfunction we have

$$\varphi_\lambda(x) \quad \sim \quad \begin{cases} \mathrm{e}^{\mathrm{i}\lambda x} & \text{for } x \ll 0 \\ \dfrac{\lambda + \mathrm{i}k}{\lambda - \mathrm{i}k}\,\mathrm{e}^{i\lambda x} & \text{for } x \gg 0\,. \end{cases}$$

Lax equations

If L_u is the operator (1.3) and[2]

$$P_u = 4\left(\frac{\mathrm{d}}{\mathrm{d}x}\right)^3 - 6u\frac{\mathrm{d}}{\mathrm{d}x} - 3u' \tag{1.4}$$

then it turns out that the commutator $[P_u, L_u]$, which would be expected to be a differential operator, actually reduces to the zero-order operator of multiplication by

$$u''' - 6uu'\,.$$

The KdV equation, which is an evolution equation for u, can therefore be expressed as an evolution equation

$$\frac{\mathrm{d}}{\mathrm{d}t}L_u = [P_u, L_u] \tag{1.5}$$

for the operator L_u. An equation of this form is called a *Lax equation.* Its significance is that it describes an *isospectral* evolution of the operator L_u. For if u depends on a parameter t, and we find g_t so that $g_0 = 1$ and

$$\frac{\mathrm{d}}{\mathrm{d}t}\, g_t = P_u\, g_t\,,$$

[2]Where P_u comes from will be explained in Section 4.

then (1.5) is equivalent to

$$\frac{\mathrm{d}}{\mathrm{d}t}\{g_t^{-1}L_u\,g_t\}=0\,,$$

so that

$$L_u(t)=g_t\,L_u(0)\,g_t^{-1}\,.$$

The significance of the Lax form for solving the KdV equation is that if we let the eigenfunction φ_λ evolve in time according to

$$\frac{\partial\varphi_\lambda}{\partial t}=P_u\varphi_\lambda$$

then it will remain an eigenfunction for L_u when u evolves according to the KdV equation. Furthermore, because P_u reduces simply to $4(\mathrm{d}/\mathrm{d}x)^3$ when $|x|$ is large, the evolving φ_λ will have the form

$$\varphi_\lambda(x,t)=\begin{cases}\mathrm{e}^{\mathrm{i}\lambda x-4\mathrm{i}\lambda^3 t}+R_\lambda\mathrm{e}^{-\mathrm{i}\lambda x+4\mathrm{i}\lambda^3 t} & \text{for } x\ll 0\\ T_\lambda\mathrm{e}^{\mathrm{i}\lambda x-4\mathrm{i}\lambda^3 t} & \text{for } x\gg 0\,.\end{cases}$$

This means that the transmission coefficient T_λ remains constant in time, while the reflection coefficient R_λ evolves according to the simple law

$$R_\lambda(t)=\mathrm{e}^{8\mathrm{i}\lambda^3 t}R_\lambda(0)\,.$$

Similarly, when there are solitons present, parametrized by (ω_k,ρ_k), then ω_k remains constant, and ρ_k evolves by

$$\rho_k(t)=\mathrm{e}^{-8\omega_k^3 t}\rho_k(0)\,.$$

If we have a way of recovering the function u from its scattering data, therefore, we have a complete solution to the KdV equation, precisely analogous to the solution of a linear evolution equation by means of the spatial Fourier transform.

Two other aspects of the situation should be mentioned in this introduction. For each λ the transmission coefficient T_λ remains constant under the KdV flow, and so we have an uncountable collection of conservation laws. But T_λ depends in a complicated global way on the function u, and we should like to have *local* conservation laws, i.e. we should like the conserved quantity to be of the form

$$\int_{\mathbb{R}}F(u,u',u'',\ldots)\,\mathrm{d}x,$$

where F is a polynomial in u and its x-derivatives. Such a function F is called a *conserved density*. Two obvious conserved quantities of this form are

$$\int_{\mathbb{R}} u\,\mathrm{d}x \qquad \text{and} \qquad \int_{\mathbb{R}} \left\{\tfrac{1}{2}(u')^2 - u^3\right\} \mathrm{d}x\,.$$

A complete sequence of local conservation laws can be obtained from the asymptotic expansion of $\log T_\lambda$ as $\lambda \to \infty$. We have

$$\log T_\lambda \sim \sum_{k \geqslant 0} \frac{I_k}{\lambda^{k+1}}$$

as $\lambda \to \infty$, where each I_k is the integral of a local conserved density of the above form. The reason for this localization will be discussed in Section 10.

The second point to be made is that T_λ can be regarded as the 'characteristic polynomial' $\det(\lambda^2 - L_u)$ of the operator L_u. In isospectral flows the characteristic polynomial is always the natural invariant to look at. The relation between T_λ and $\det(\lambda^2 - L_u)$ is the subject of Section 9.

2 Classical dynamical systems and integrability

A state of a dynamical system was described classically by n 'position variables' $x_1, \ldots, x_n$, and n 'momentum variables' $p_1, \ldots, p_n$. Its evolution was determined by a single function

$$H(x_1, \ldots, x_n;\, p_1, \ldots, p_n)\,,$$

called the *Hamiltonian*, by means of *Hamilton's equations*

$$\frac{\mathrm{d}x_i}{\mathrm{d}t} = \frac{\partial H}{\partial p_i}, \quad \frac{\mathrm{d}p_i}{\mathrm{d}t} = -\frac{\partial H}{\partial x_i}\,. \tag{2.1}$$

Thus the smooth function H, defined, say, in an open set U of $\mathbb{R}^{2n}$, gives us a vector field

$$\xi_H = \sum \frac{\partial H}{\partial p_i}\frac{\partial}{\partial x_i} - \sum \frac{\partial H}{\partial x_i}\frac{\partial}{\partial p_i} \tag{2.2}$$

on U according to which the system evolves.

If $F : U \to \mathbb{R}$ is another function, we can define a vector field ξ_F by replacing H by F in (2.1) or (2.2). The rate of change of F along a trajectory of (2.1) is given by the *Poisson bracket*

$$\{H, F\} = D_{\xi_H} F = -D_{\xi_F} H\,. \tag{2.3}$$

(Here D_{ξ_H} is the directional derivative along the vector field ξ_H.)

In particular, F is 'conserved', i.e. constant, on trajectories of (2.1) if and only if H is constant on trajectories of ξ_F.

Example For a particle of unit mass moving in $\mathbb{R}^3$ under the force produced by a gravitational potential $V : \mathbb{R}^3 \to \mathbb{R}$ we have $n = 3$, and

$$H = \tfrac{1}{2}(p_1^2 + p_2^2 + p_3^2) \; + \; V(x_1, x_2, x_3)\,,$$

leading to the equations of motion

$$\dot{x}_i = p_i, \quad \dot{p}_i = -\frac{\partial V}{\partial x_i}\,. \tag{2.4}$$

If $F = x_2 p_3 - x_3 p_2$ the vector field

$$\xi_F = x_2 \frac{\partial}{\partial x_3} - x_3 \frac{\partial}{\partial x_2}$$

describes rotation of $\mathbb{R}^3$ about the x_1-axis, and $\xi_F H = 0$ if V is unchanged by this rotation. Thus F, which is the angular momentum about the x_1-axis, is conserved by the motion (2.4) if and only if V is axially symmetric.

Obviously the simplest dynamical system would be one for which H depends only on $p_1, \ldots, p_n$, for then Hamilton's equations tell us that the p_i are constants of the motion, while the x_i evolve linearly in time with speed

$$\dot{x}_i = \frac{\partial H}{\partial p_i}\,.$$

Traditionally, a system is called *integrable* if it takes this form when 'written in suitable coordinates'. But what kind of coordinate changes are allowed? It is hard to make sense of this question in a single coordinate patch. Hamilton's equations can be expressed a little more intrinsically

$$\xi_H(u) = K_u \cdot \mathrm{d}H(u)$$

where $\xi_H(u)$ is the value at $u \in U$ of the vector field ξ_H associated to H, while

$$\mathrm{d}H(u) \in T_u^*U$$

is the gradient of H at u, and

$$K_u : T_u^*U \to T_uU$$

is a skew linear map taking cotangent vectors to tangent vectors at u, which in terms of the bases

$$\{\mathrm{d}x_i, \mathrm{d}p_i\}, \qquad \left\{\frac{\partial}{\partial x_i}, \frac{\partial}{\partial p_i}\right\}$$

is given by the $2n \times 2n$ matrix

$$K_u = \begin{pmatrix} 0 & 1 \\ -1 & 0 \end{pmatrix}. \tag{2.5}$$

We can think of K_u as an element of $\wedge^2 T_uU$. Thus the formalism of classical mechanics makes sense whenever we have a manifold U with a skew tensor field K, i.e. a section of $\wedge^2 TU$.

Definition 2.6 *A Poisson manifold is a manifold U equipped with a skew tensor field K such that the associated Poisson bracket on functions on U satisfies the Jacobi identity*

$$\{f,\{g,h\}\} + \{g,\{h,f\}\} + \{h,\{f,g\}\} = 0.$$

A symplectic manifold is a Poisson manifold for which K_u is invertible at each point $u \in U$.

Remark On a symplectic manifold K^{-1} is a 2-form ω, and the Jacobi identity holds if and only if ω is *closed.*

For a symplectic manifold, *Darboux's theorem* tells us that locally one can always find a coordinate system $\{x_1, \ldots, x_n;\ p_1, \ldots, p_n\}$ in which K takes the standard form (2.5), i.e.

$$\omega = K^{-1} = \sum \mathrm{d}p_i \wedge \mathrm{d}x_i\,.$$

Such coordinates are called *canonical.* Furthermore, it is easy to show that if $p_1, \ldots, p_n : U \to \mathbb{R}$ are functions on a $2n$-dimensional symplectic manifold which are independent near $u \in U$ and 'in involution', i.e. such that the Poisson brackets $\{p_i, p_j\}$ all vanish, then one can find functions $x_1, \ldots, x_n$ defined near u such that $\{x_1, \ldots, p_n\}$ is a canonical coordinate system. In this situation the functions p_i and x_i are called 'action' and 'angle' variables respectively.

We can now attempt a definition.

Definition 2.7 *A $2n$-dimensional symplectic manifold U with a Hamiltonian $H : U \to \mathbb{R}$ is* integrable *if there exist n conserved functions $p_i : U \to \mathbb{R}$ which are in involution and are independent at almost all points of U.*

The Hamiltonian will then be a function only of the p_i, for if we introduce canonical coordinates in the neighbourhood of a point where the p_i are independent we shall have

$$\frac{\partial H}{\partial x_i} = -\{p_i, H\} = 0$$

as p_i is conserved.

When we have an integrable system the subsets

$$U_\eta = p^{-1}(\eta) = \{u \in U : p_i(u) = \eta_i\}$$

are manifolds for almost all $\eta \in \mathbb{R}^n$, and they are the orbits of the action of the additive group $\mathbb{R}^n$ generated by the vector fields ξ_{p_i}. An important case is when the surfaces of constant H are *compact.* Then each submanifold U_η is compact, and must necessarily be a *torus* $\mathbb{R}^n/L$, where $L \cong \mathbb{Z}^n$ is a lattice in $\mathbb{R}^n$. This is why the coordinates x_i on U_η are called 'angle' variables. Generically, a trajectory of the system will wind densely around one of the tori U_η.

An integrable system is the opposite of a chaotic system. To see the difference, consider geodesic motion on a compact surface Σ with a Riemannian metric. Then U is the tangent bundle $T\Sigma$, and the Hamiltonian is $H(x, \dot{x}) = \frac{1}{2}||\dot{x}||^2$. If Σ is an ellipsoid in $\mathbb{R}^3$ the system is integrable (see Section 3), and each trajectory winds round a two-dimensional torus in $T\Sigma$. These tori project to 'belts' on the ellipsoid

Σ, and if x belongs to the belt of a geodesic then the geodesic will continually pass close to x, and each time will be going in the same direction. If Σ is a surface of genus > 1 with a metric of negative curvature, on the other hand, then the geodesic motion is chaotic, and a generic trajectory is dense in the three-dimensional compact manifold $||\dot{x}||^2 =$ constant. A generic geodesic will continually pass close to any given point $x \in \Sigma$, but is equally likely to be going in any direction when it does so.

A feature of many integrable systems is that each torus U_η on which the system evolves is the Jacobian torus of an algebraic curve M_η which depends on the value of η of the action variables. One way in which that comes about is as follows. For the systems we study the phase-space U is often a space of 'operators' L of some kind, and the 'action variables' are the coefficients of the characteristic polynomial of L. If there is another operator Q_η which commutes with all L in U_η then we can consider the *joint spectrum* of L and Q_η, i.e. the set of pairs (λ, μ) such that

$$\begin{aligned} L\psi &= \lambda\psi \\ Q_\eta\psi &= \mu\psi \end{aligned} \tag{2.8}$$

for some non-zero ψ. Such pairs (λ, μ) commonly form an algebraic curve M_η, and for each $L \in U_\eta$ the solutions ψ of (2.8) form a line $E_{L;\lambda,\mu}$, i.e. a point of the Jacobian of M_μ.

Example A beautiful infinite-dimensional example of the archetypal situation has been worked out by McKean and Trubowitz (1976). In the case of the KdV flow on periodic functions u on the line—say with period 2π—we have L_u as in Section 1, and can take Q to be the operation of translation by 2π. Then Q acts on each two-dimensional space V_λ of solutions of $L_u\psi = \lambda^2\psi$ with characteristic polynomial

$$\det(\mu - Q_\lambda) = \mu^2 - F(\lambda)\mu + 1,$$

where $Q_\lambda = Q|V_\lambda$, and $F(\lambda) = \mathrm{tr}(Q_\lambda)$. The function $F(\lambda)$ is essentially the characteristic polynomial of L_u on periodic functions (see Section 9). If F were a polynomial then

$$\det(\mu - Q_\lambda) = 0$$

would be the equation of a hyperelliptic curve M. In McKean and Trubowitz (1976) it is shown that in any case M is a 'hyperelliptic curve of possibly infinite genus'.

A line bundle E on M is given concretely by the sets of zeros and poles of a holomorphic section ψ of E. We choose a definite section ψ by prescribing $\psi(0) = 1$. This never vanishes, but blows up when (2.8) implies that $\psi(0) = \psi(2\pi) = 0$, i.e. when λ is an eigenvalue of the Dirichlet problem for L_u on $[0, 2\pi]$. Thus the 'angle variables' are the spectrum of the Dirichlet problem. In McKean and Trubowitz (1976) it is shown that the set of L_u with given periodic spectrum (i.e. a given function $F(\lambda)$) is a torus, generically of infinite dimension, and that the space U of periodic functions u is foliated by these tori.

3 Some classical integrable systems

In 1838 Jacobi proved that the equations of geodesic motion on an ellipsoid Σ in $\mathbb{R}^{n+1}$ were integrable. The theory of this system is very attractively presented by Moser (1979). It is a little too complicated to describe here, so I shall just say that the equations can be written in Lax form $\dot{L} = [P, L]$, describing isospectral deformations of a matrix L; and the n conserved quantities in involution which provide the integrability are the coefficients of the characteristic polynomial $\det(\lambda - L)$. To give the Lax form, we write the equation of the ellipsoid Σ as

$$x^t A x = 1\,, \tag{3.1}$$

where A is a positive definite symmetric $(n+1) \times (n+1)$ matrix. A possible velocity of a particle on Σ is then a pair $(x, p = \dot{x})$ satisfying (3.1) and also $x^t A p = 0$. To (x, p) we associate the matrix

$$L = \Pi_p (A^{-1} - x x^t) \Pi_p\,,$$

where $\Pi_p = 1 - ||p||^{-2} p p^t$ is the matrix representing orthogonal projection on to the hyperplane orthogonal to p.

A rather similar system which is simpler to treat is the *finite Toda lattice.* Here we have n numbered particles of unit mass on a line, with positions $x_1, \ldots, x_n$, moving under the action of a force derived from the potential

$$V(x_1, \ldots, x_n) = \sum_{i=1}^{n=1} \mathrm{e}^{2k(x_i - x_{i+1})},$$

where k is a given positive constant. The Hamiltonian is

$$H = \tfrac{1}{2}\Sigma p_i^2 + V(x_1, \ldots, x_n),$$

where $p_i = \dot{x}_i$. As $k \to \infty$ this becomes the problem of n elastic billiard balls at positions $x_1 < x_2 < \cdots < x_n$, moving freely except for collisions in which the colliding balls exchange velocities. For the billiard balls the elementary symmetric functions $\sigma_1, \ldots, \sigma_n$ of the velocities $p_1, \ldots, p_n$ are conserved quantities in involution, and the Hamiltonian is

$$\tfrac{1}{2}\Sigma p_i^2 = \tfrac{1}{2}\sigma_1^2 - \sigma_2.$$

For any value of k we define the symmetric tri-diagonal matrix

$$L = \begin{pmatrix} p_1 & b_1 & 0 & 0 & \cdot & \cdot & \cdot \\ b_1 & p_2 & b_2 & 0 & \cdot & \cdot & \cdot \\ 0 & b_2 & p_3 & b_3 & \cdot & \cdot & \cdot \\ & \cdots & & & & \cdots & \\ & \cdots & & & & \cdots & \\ & \cdots & & & b_{n-2} & p_{n-1} & b_{n-1} \\ & \cdots & & & 0 & b_{n-1} & p_n \end{pmatrix},$$

where $b_k = e^{k(x_k - x_{k+1})}$, and the skew matrix

$$P = \begin{pmatrix} 0 & -b_1 & 0 & \cdot & \cdot & \cdot & 0 \\ b_1 & 0 & -b_2 & \cdot & \cdot & \cdot & 0 \\ 0 & b_2 & 0 & \cdot & \cdot & \cdot & 0 \\ & \cdots & & & & & \\ & \cdots & & & & & \\ & \cdots & & & 0 & 0 & -b_{n-1} \\ & \cdots & & & 0 & b_{n-1} & 0 \end{pmatrix}.$$

Then the equations of motion are $\dot{L} = [P, L]$, and the conserved quantities in involution are the coefficients of the characteristic polynomial $\det(\lambda - L)$. The Hamiltonian is $\tfrac{1}{2}\mathrm{tr}(L^2)$. If the particles are widely separated then the b_i are very small, and $\det(\lambda - L)$ is very close to $\prod(\lambda - p_i)$, i.e. the conserved quantities are very nearly the elementary symmetric functions of the p_i, just as in the case of elastic balls. This system is described in detail in Moser (1975).

Euler equations

One of the earliest systems known to be integrable was the free motion of a rigid body pivoted at its centre of mass. The positions of the body form the group manifold $G = SO_3$, and the Hamiltonian is the quadratic form $H : TG \to \mathbb{R}$ on tangent vectors given by a left-invariant Riemannian metric on G. (This quadratic form, which is determined by its value on the tangent space at the identity-element of G, i.e. on the Lie algebra $\mathfrak{g}$ of G, is called the *inertia tensor* of the rigid body.) A typical geodesic for a left-invariant metric on SO_3 winds densely around a two-dimensional rather than a three-dimensional torus. A more typical integrable system is the motion of a 'top' with an axis of symmetry which is pivoted at a point on its axis—not the centre of mass—and moves under gravity. Then the generic orbit is dense on a three-dimensional torus, for the top spins about its axis, while the axis 'precesses' steadily around the vertical, and at the same time 'nutates'—i.e. the angle between the axis and the vertical oscillates.

Geodesic motion for a left-invariant metric on an arbitrary Lie group is known to be integrable for a certain class of metrics, but not all. I shall not discuss that, but shall describe a general method of constructing conserved quantities in involution for such systems.

The *angular velocity* of a rigid body whose position is $g \in G$ is the element $\omega = g^{-1}\dot{g}$ of the Lie algebra $\mathfrak{g}$. Its *angular momentum* is $\alpha = H(\omega)$ in the dual space $\mathfrak{g}^*$, where H is now regarded as an isomorphism $H : \mathfrak{g} \to \mathfrak{g}^*$. The equations for a geodesic can be written as a Lax equation for $\alpha \in \mathfrak{g}^*$, namely

$$\frac{\mathrm{d}}{\mathrm{d}t}\,\alpha = [\omega, \alpha] = [H^{-1}(\alpha), \alpha]\,. \tag{3.2}$$

This is called *Euler's equation.* (If α is known, then so is ω, and then g can be found, in principle, by integration.) The equation describes geodesic motion for an arbitrary left-invariant metric on any group.

Euler's equation is a vector field on $\mathfrak{g}^*$ associated to the quadratic form H^{-1} on $\mathfrak{g}^*$. The manifold $\mathfrak{g}^*$ is, in fact, a *Poisson manifold* (cf. Definition 2.6): for any smooth $F : \mathfrak{g}^* \to \mathbb{R}$ we have a vector field D_F on $\mathfrak{g}^*$ whose action $D_F G = \{F, G\}$ on a function $G : \mathfrak{g}^* \to \mathbb{R}$ is given by

$$\{F, G\}(\alpha) \;=\; \alpha([dF(\alpha), dG(\alpha)])\,,$$

for $\alpha \in \mathfrak{g}^*$. (Here $dF(\alpha)$ and $dG(\alpha)$ are elements in $(\mathfrak{g}^*)^* = \mathfrak{g}$, and $[\ ,\ ]$ is the bracket in $\mathfrak{g}$.)

There is a standard way to produce a family of Poisson-commuting functions on $\mathfrak{g}^*$. If $F : \mathfrak{g}^* \to \mathbb{R}$ is any G-invariant function, and $\beta \in \mathfrak{g}$ is an arbitrary fixed element, then for any $\lambda \in \mathbb{R}$ let $F_\lambda : \mathfrak{g}^* \to \mathbb{R}$ be defined by

$$F_\lambda(\alpha) = F(\alpha - \lambda\beta)\,.$$

Proposition 3.3 *We have $\{F_\lambda, F_\mu\} = 0$ for any λ, μ.*

Proof The G-invariance of F is expressed by

$$(\xi \cdot \alpha)(\mathrm{d}F(\alpha)) = 0,$$

for every $\alpha \in \mathfrak{g}^*$ and $\xi \in \mathfrak{g}$, or equivalently

$$\alpha([\xi, \mathrm{d}F(\alpha)] = 0\,.$$

If in this we replace α by $\alpha - \mu\beta$ and ξ by $\mathrm{d}F(\alpha - \lambda\beta)$, we get

$$(\alpha - \mu\beta)([\mathrm{d}F(\alpha - \lambda\beta),\ \mathrm{d}F(\alpha - \mu\beta)]) = 0\,.$$

Similarly

$$(\alpha - \lambda\beta)([\mathrm{d}F(\alpha - \lambda\beta),\ \mathrm{d}F(\alpha - \mu\beta)]) = 0\,.$$

But these two equations imply

$$\alpha([\mathrm{d}F(\alpha - \lambda\beta),\ \mathrm{d}F(\alpha - \mu\beta)]) = 0,$$

which is what we want to prove. □

4 Formal pseudo-differential operators

The Lax form $\mathrm{d}L/\mathrm{d}t = [P, L]$ of the KdV equation depends upon the existence of a third-order differential operator P whose commutator with

$$L = -\left(\frac{\mathrm{d}}{\mathrm{d}x}\right)^2 + u$$

is a multiplication-operator. The most practical way to find such operators P is by the calculus of formal pseudo-differential operators. Pseudo-differential operators are the smallest class of operators which includes both differential operators and their inverses. Their definition is rather complicated, and need not concern us, for in the one-dimensional case a formal algebraic treatment is all we need.[3] What follows is due to Gelfand and Dikii (1976), but the present elegant treatment is taken from Wilson (1979).

Let us consider formal power series

$$f_n D^n + f_{n-1} D^{n-1} + \cdots + f_0 + f_{-1} D^{-1} + f_{-2} D^{-2} + \cdots \qquad (4.1)$$

in an indeterminate D (to be thought of as $-\mathrm{i}d/dx$), whose coefficients f_i belong to a commutative ring $\mathcal{F}$ (thought of as the scalar-valued functions on some interval in $\mathbb{R}$). We assume $\mathcal{F}$ is equipped with a *derivation*, written $f \mapsto f'$. We multiply such formal series using the rule

$$Df = fD + f',$$

which implies

$$D^{-1} f = f D^{-1} - f' D^{-2} + f'' D^{-3} - \cdots,$$

and, in general,

$$D^k f = \sum_{r \geqslant 0} \binom{k}{r} f^{(r)} D^{k-r},$$

where

$$\binom{k}{r} = \frac{1}{r!} k(k-1) \cdots (k-r+1).$$

This gives us a non-commutative ring $\mathcal{D}$ containing $\mathcal{F}$.

If P denotes the operator (4.1) we write P_+ for its 'differential operator part'

$$P_+ = f_n D^n + \cdots + f_1 D + f_0,$$

and

$$P_- = f_{-1} D^{-1} + f_{-2} D^{-2} + \cdots,$$

so that

$$P = P_+ + P_-.$$

The element $f_{-1} \in \mathcal{F}$ will be called the *residue* $\mathrm{res}(P)$ of P.

---

[3]But see the remarks on page 70.

For any $u \in \mathcal{F}$ the element D^2+u of $\mathcal{D}$ has a unique square-root $L^{\frac{1}{2}}$ of the form

$$L^{\frac{1}{2}} = D + \tfrac{1}{2}uD^{-1} - \tfrac{1}{4}u'D^{-2} + \tfrac{1}{8}(u'' - u^2)D^{-3} + \cdots .$$

For any other positive integer k we can therefore define the 'fractional power'

$$L^{\frac{k}{2}} = D^k + \frac{k}{2}uD^{k-2} + \frac{1}{4}k(k-2)u'D^{k-3} + \cdots .$$

Our task of finding Lax pairs is now solved by

Proposition 4.2 *For every $k > 0$ the commutator*

$$[(L^{\frac{k}{2}})_+ \, , L]$$

is a multiplication operator, i.e. an element of $\mathcal{F}$.

Proof Evidently $[(L^{\frac{k}{2}})_+ \, , L]$ involves only positive powers of D. On the other hand, $L^{\frac{k}{2}}$ commutes with $L = (L^{\frac{1}{2}})^2$, so

$$[(L^{\frac{k}{2}})_+ \, , L] = -[(L^{\frac{k}{2}})_- \, , L] .$$

The right-hand side is of the form

$$-[fD^{-1} + \cdots , D^2 + u] = 2f' + (\text{ lower terms}) .$$

So we must have $[(L^{\frac{k}{2}})_+ \, , L] = 2f'$. □

In view of this proposition we can define a flow ∂_k on the operators L for each $k \geqslant 1$ by $\partial_k L = [P_k^+, L]$, where $P_k^+ = (L^{k/2})_+$. (When k is even the flow is zero, as P_k^+ is a power of L.)

Proposition 4.3 *The flows ∂_k and ∂_m commute for all k, m. Furthermore, the residue* $\mathrm{res}(P_m^+)$ *is a conserved density of the flow ∂_k.*

The last statement means that $\partial_k \mathrm{res}(P_m^+) = f'_{km}$ where $f_{km} \in \mathcal{F}$ is a polynomial in $u, u', u'', \ldots$. This is a local substitute for

$$\partial_k \int \mathrm{res}(P_m^+) \, \mathrm{d}x = 0 ,$$

which follows from it when appropriate boundary conditions are satisfied (e.g. if u has compact support on $\mathbb{R}$, or else is periodic and the integral is over a period).

To give the proof it is best to go one step further in the direction of algebraization. So far we have made no assumptions about the differential algebra $\mathcal{F}$ which contains the coefficient u of the operator $L = D^2 + u$, but from now on we shall take $\mathcal{F}$ to be the polynomial algebra $\mathbb{C}[u, u', u'', \ldots]$ on a sequence of formal indeterminates—i.e. to be the *free* differential algebra generated by u. Then the formula

$$\partial_k u = \partial_k L = [P_k^+, L] \in \mathcal{F}$$

defines a unique derivation $\partial_k : \mathcal{F} \to \mathcal{F}$ such that $\partial_k(f') = (\partial_k f)'$. We extend ∂_k coefficient-wise to a derivation $\partial_k : \mathcal{D} \to \mathcal{D}$, i.e. so that $\partial_k D = 0$.

Then we have

$$\begin{aligned} \partial_k \partial_m L &= \partial_k [P_m^+, L] \\ &= [\partial_k P_m^+, L] + [P_m^+, [P_k^+, L]]\,. \end{aligned}$$

So

$$[\partial_k, \partial_m] L = [F_{km}, L]\,,$$

where

$$F_{km} = \partial_k P_m^+ - \partial_m P_k^+ - [P_k^+, P_m^+]\,.$$

But

$$\begin{aligned} \partial_k P_m^+ &= (\partial_k L^{m/2})_+ \\ &= [P_k^+, L^{m/2}]_+ \\ &= -[P_k^-, L^{m/2}]_+ \\ &= -[P_k^-, P_m^+]_+\,. \end{aligned}$$

So

$$\begin{aligned} F_{km} &= -[P_k^-, P_m^+]_+ - [P_k^+, P_m^-]_+ - [P_k^+, P_m^+] \\ &= [P_k^-, P_m^-]_+ \\ &= 0\,. \end{aligned}$$

In this calculation we have used only the obvious fact that $L^{k/2} = P_k^+ + P_k^-$ commutes with $L^{m/2} = P_m^+ + P_m^-$, and in addition the formula

$$\partial_k L^{m/2} = [P_k^+, L^{m/2}],$$

which—as both ∂_k and $[P_k^+,\ \ ]$ are derivations—needs to be proved only when $m = 1$. But $\partial_k L = [P_k^+, L]$, so

$$\partial_k L^{1/2} - [P_k^+, L^{1/2}]$$

must anticommute with $L^{1/2} = D + \cdots$, from which it is easily seen to vanish.

The conservation of the residue is an even easier calculation.

$$\begin{aligned} \partial_k \operatorname{res}(P_m) &= \operatorname{res}(\partial_k P_m) \\ &= \operatorname{res}[P_k^+, P_m]. \end{aligned}$$

But in the algebra $\mathcal{D}$ the residue of every commutator is a derivative, for

$$\operatorname{res}\, [fD^p, gD^q] = \binom{p}{p+q+1} (fg^{(p+q+1)} - (-1)^{p+q+1} f^{(p+q+1)} g),$$

and

$$fg^{(n)} - (-1)^n f^{(n)} g = \{\sum (-1)^i f^{(i)} g^{(n-i-1)}\}'.$$

Genuine operators

It is natural to ask what the formal algebra $\mathcal{D}$ has to do with genuine pseudo-differential operators. In fact $\mathcal{D}$ is the algebra of *symbols* of pseudo-differential operators: every operator has a symbol, which determines it up to a *smoothing operator*, i.e. an integral operator whose kernel is smooth in both variables. The pseudo-differential operators form an algebra $\mathcal{P}$ which operates on the ring of smooth functions with compact support on the interval I. There is a ring-homomorphism $\mathcal{P} \to \mathcal{D}$ whose kernel (in the other sense!) is the ideal of smoothing operators. An operator has the symbol

$$f_{-1}D^{-1} + f_{-2}D^{-2} + \cdots$$

if it is an integral operator with a kernel $K(x, y)$ with proper support (i.e. $k(x,\ \ )$ and $k(\ \ , y)$ have compact support for each x, y) which is smooth except for a jump discontinuity on the diagonal $x = y$, where $(-\mathrm{i}\partial/\partial x)^m k(x,y)$ jumps by f_{-m-1}.

Actually, even this is an overstatement: the true algebra of pseudo-differential operators on I is $\mathcal{P} + \sigma\mathcal{P}$, where $\sigma = D/|D|$ is a singular integral operator whose kernel looks like $(x-y)^{-1}$ on the diagonal. We have $\sigma^2 = 1$ and $\sigma D = D\sigma$ modulo smoothing operators.

5 Scattering theory

We shall now consider in a little more detail the scattering theory associated with the operator

$$L_u = -\left(\frac{\mathrm{d}}{\mathrm{d}x}\right)^2 + u\,,$$

where u is a smooth real-valued function with compact support on the line. A good reference for the following material is Faddeev and Takhtajan (1987).

For any λ the equation $L_u\varphi = \lambda^2\varphi$ has a two-dimensional solution space V_λ. Let φ_λ^+ and φ_λ^- be the solutions such that

$$\left.\begin{aligned} \varphi_\lambda^+(x) &= \mathrm{e}^{\mathrm{i}\lambda x} \\ \varphi_\lambda^-(x) &= \mathrm{e}^{-\mathrm{i}\lambda x} \end{aligned}\right\} \quad \text{when } x \ll 0\,,$$

and let $\tilde{\varphi}_\lambda^\pm$ be the two solutions such that

$$\tilde{\varphi}_\lambda^\pm(x) = \mathrm{e}^{\pm\mathrm{i}\lambda x} \quad \text{when } x \gg 0\,.$$

These two bases for V_λ must be related:[4]

$$\begin{aligned} \tilde{\varphi}_\lambda^+ &= a_\lambda\varphi_\lambda^+ + b_\lambda\varphi_\lambda^- \\ \tilde{\varphi}_\lambda^- &= c_\lambda\varphi_\lambda^+ + d_\lambda\varphi_\lambda^-\,, \end{aligned}$$

where the 2×2 matrix

$$g_\lambda = \begin{pmatrix} a_\lambda & b_\lambda \\ c_\lambda & d_\lambda \end{pmatrix}$$

[4] Of course, $\varphi_\lambda^\pm$ and $\tilde{\varphi}_\lambda^\pm$ are not bases when $\lambda = 0$, and g_λ as I have defined it has a pole at $\lambda = 0$, which can be removed by using the basis $\{\cos\lambda x,\ \lambda^{-1}\sin\lambda x\}$ instead of $\mathrm{e}^{\pm\mathrm{i}\lambda x}$. In the following discussion I shall ignore what happens when $\lambda = 0$, as it does not affect the essential argument. In fact scattering for L_u is related to a 'twisted' form of the loop group of SL_2. But cf. Section 6.

is invertible and independent of x. In fact g_λ belongs to $SL_2\mathbb{C}$, because the *Wronskian* $W(\varphi,\psi) = \varphi\psi' - \varphi'\psi$ of two solutions of $L_u\varphi = \lambda^2\varphi$ is constant. (We also have $d_\lambda = a_{-\lambda} = \bar{a}_{\bar{\lambda}}$ and $c_\lambda = b_{-\lambda} = \bar{b}_{\bar{\lambda}}$.) The transmission and reflection coefficients of Section 1 are given by

$$T_\lambda = a_\lambda^{-1} \quad \text{and} \quad R_\lambda = a_\lambda^{-1} b_\lambda .$$

As $\lambda \to \pm\infty$ along the real axis the matrix g_λ tends rapidly to 1: intuitively, a very high-frequency wave barely notices a small obstacle. Thus the scattering caused by u gives us a smooth based loop in the group $SL_2\mathbb{C}$. ('Based' means that $g_\lambda = 1$ when $\lambda = \pm\infty$.) I shall call this the *holonomy loop* of u. This is where loop groups enter the theory.

If u has compact support then $\varphi_\lambda^\pm, \tilde{\varphi}_\lambda^\pm$, and g_λ are all defined and holomorphic in λ for all $\lambda \in \mathbb{C}$. But unless $u = 0$ we do not expect g_λ to tend to 1 when $\lambda \to \infty$ in complex directions: it has an essential singularity at $\lambda = \infty$. In any case the holomorphicity of g_λ is misleading. If we assume only that u is rapidly decreasing as $x \to \pm\infty$ then for $\mathrm{Im}(\lambda) > 0$ we can define φ_λ^- as the unique solution of $L_u\varphi = \lambda^2\varphi$ which tends to zero like $\mathrm{e}^{-\mathrm{i}\lambda x}$ as $x \to -\infty$, and $\tilde{\varphi}_\lambda^+$ as the unique solution which tends to zero like $\mathrm{e}^{\mathrm{i}\lambda x}$ as $x \to +\infty$. It does not make sense, however, to try to characterize a solution φ_λ^+ by $\varphi_\lambda^+ \sim \mathrm{e}^{\mathrm{i}\lambda x}$ as $x \to -\infty$, for adding any multiple of φ_λ^- would not affect that asymptotic behaviour. In fact φ_λ^- and $\tilde{\varphi}_\lambda^+$ are defined and holomorphic in λ when $\mathrm{Im}(\lambda) > 0$, while φ_λ^+ and $\tilde{\varphi}_\lambda^-$ are defined and holomorphic when $\mathrm{Im}(\lambda) < 0$. Only when λ is real are all four solutions defined, and so g_λ is defined only for $\lambda \in \mathbb{R}$. It is still a smooth loop in $SL_2\mathbb{C}$, just as when u has compact support.

The holomorphicity properties just described—in fact the stronger assertion that for each x the functions

$$\varphi_\lambda^\pm(x)\mathrm{e}^{\mp\mathrm{i}\lambda x} \quad \text{and} \quad \tilde{\varphi}_\lambda^\pm(x)\mathrm{e}^{\mp\mathrm{i}\lambda x}$$

are holomorphic and *bounded* in their respective half-planes—are the link between the time-independent behaviour of the operator L_u and the scattering for the equation

$$\ddot{\varphi} + L_u\varphi = 0 . \tag{5.1}$$

To understand this we repeatedly use the basic fact that if f is a smooth function of rapid decrease on $\mathbb{R}$ then $f(x) = 0$ for $x \geqslant 0$ if and only if the Fourier transform

$$\hat{f}(\lambda) = \frac{1}{2\pi}\int_{\mathbb{R}} f(x)\mathrm{e}^{-\mathrm{i}\lambda x}\,\mathrm{d}x$$

extends to a bounded holomorphic function in the upper half-plane.[5] It follows from this that f has support in $[-R, R]$ if and only if $\hat{f}$ is entire and

$$|\hat{f}(\lambda)| \leqslant C\mathrm{e}^{R|\mathrm{Im}(\lambda)|}. \tag{5.2}$$

The standard way to write a solution of (5.1) is

$$\varphi(x,t) = \int_{\mathbb{R}} \varphi_\lambda(x)\mathrm{e}^{\mathrm{i}\lambda t}\,\mathrm{d}\lambda, \tag{5.3}$$

where φ_λ belongs to the eigenspace V_λ of L_u. Using the basis $\{\varphi_\lambda^+, \tilde{\varphi}_\lambda^-\}$ of V_λ we have

Proposition 5.4 *If f_1 and f_2 have compact support on $\mathbb{R}$, and $\varphi(x,t)$ is defined by* (5.3) *with*

$$\varphi_\lambda(x) = \hat{f}_1(\lambda)\varphi_\lambda^+(x) + \hat{f}_2(\lambda)\tilde{\varphi}_\lambda^-(x),$$

then

$$\varphi(x,t) = f_1(t+x) + f_2(t-x)$$

for $t \ll 0$, i.e. φ consists of two incoming wave-packets approaching the obstacle.

Proof We have

$$\begin{aligned}\varphi(x,t) &= \int \{\varphi_\lambda^+(x)\mathrm{e}^{-\mathrm{i}\lambda x}\hat{f}_1(\lambda)\}\mathrm{e}^{\mathrm{i}\lambda(t+x)}\,\mathrm{d}\lambda \\ &\qquad + \int \{\tilde{\varphi}_\lambda^-(x)\mathrm{e}^{\mathrm{i}\lambda x}\hat{f}_2(\lambda)\}\mathrm{e}^{\mathrm{i}\lambda(t-x)}\,\mathrm{d}\lambda\,.\end{aligned}$$

Because $\varphi_\lambda^+(x)\mathrm{e}^{-\mathrm{i}\lambda x}$ is holomorphic and bounded for $\mathrm{Im}(\lambda) < 0$ the first integral vanishes unless $t + x \geqslant -R$ for some R, i.e. unless $x \geqslant -t - R$, while the second integral vanishes unless $x \leqslant t + R$. But if $t \ll 0$ we have $\varphi_\lambda^+ = \mathrm{e}^{\mathrm{i}\lambda x}$ and $\tilde{\varphi}_\lambda^- = \mathrm{e}^{-\mathrm{i}\lambda x}$ in these regions, giving us the result. □

[5] 'If' by applying Cauchy's theorem to a large semicircle around the upper half-plane, and 'only if' by the Fourier inversion theorem.

To see what happens to the wave-packets for $t \gg 0$ we change basis in each V_λ to $\{\tilde{\varphi}_\lambda^+, \varphi_\lambda^-\}$. The matrix that does this is the *scattering matrix* S_λ, which is a rearrangement of the holonomy matrix g_λ:

$$S_\lambda = d_\lambda^{-1} \begin{pmatrix} 1 & b_\lambda \\ -c_\lambda & 1 \end{pmatrix} = \begin{pmatrix} T_\lambda & R_\lambda \\ -R_\lambda & T_\lambda \end{pmatrix}.$$

The scattering matrix is a loop in $GL_2\mathbb{C}$. (For real λ we have $d_\lambda = \bar{a}_\lambda = a_{-\lambda}$, and $c_\lambda = \bar{b}_\lambda$, as well as $|a_\lambda|^2 - |b_\lambda|^2 = 1$. So $\det(S_\lambda) = a_\lambda/\bar{a}_\lambda$ has modulus 1, and S_λ is unitary.)

Proposition 5.5 *If u has compact support, then for $t \gg 0$ the solution $\varphi(x,t)$ given by* 5.3 *and Proposition* (5.4) *is*

$$\varphi(x,t) = f_3(t+x) + f_4(t-x) + \text{(bound states)},$$

where

$$\begin{pmatrix} \hat{f}_3 \\ \hat{f}_4 \end{pmatrix} = S_\lambda \begin{pmatrix} \hat{f}_1 \\ \hat{f}_2 \end{pmatrix}.$$

Proof By the argument of Proposition 5.4 this would be true, without any bound states, if the entries of the matrix S_λ were holomorphic in the upper half-plane and satisfied an estimate of the form (5.2). Now S_λ is formed from g_λ, whose entries are entire functions if u has compact support. Furthermore, b_λ and c_λ satisfy estimates of type (5.2), while $a_\lambda = d_{-\lambda} \to 1$ as $\lambda \to \infty$. Thus the entries of S_λ have the desired behaviour except for poles at the zeros of a_λ in the lower half-plane. These occur when φ_λ^- is a multiple of $\tilde{\varphi}_\lambda^+$, i.e. when $\mathrm{i}\lambda$ is the frequency of a bound state.

Reviewing the proof that f has support in $(-\infty, 0]$ if $\hat{f}$ is holomorphic and bounded in the upper half-plane, we see that if $\hat{f}$ is allowed to have a finite number of poles at points λ_k in the upper half-plane, with residues ρ_k, then

$$f(x) = \sum \rho_k \mathrm{e}^{\mathrm{i}\lambda_k x}$$

for $x \geqslant 0$. Modifying accordingly the argument of Proposition 5.4 now gives us Proposition 5.5. □

The last two propositions are best understood in terms of the spectral decomposition of the unbounded operator L_u acting on the Hilbert space $\mathcal{H}$ of L^2 functions on the line.

The spectrum of L_u consists of the positive real axis together with a finite number of points $-\omega_k^2$ on the negative real axis corresponding to 'bound states' $\psi_k \in \mathcal{H}$ such that $L_u\psi_k = -\omega_k^2\psi_k$. There is an orthogonal decomposition of $\mathcal{H}$ into L_u-invariant subspaces

$$\mathcal{H} = \mathcal{H}_0 \oplus \mathcal{H}_1 ,$$

where $\mathcal{H}_1$ is the finite-dimensional space spanned by the ψ_k, and $\mathcal{H}_0$ corresponds to the continuous spectrum $\mathbb{R}_+$, which has multiplicity 2, being spanned by φ_λ^- and $\tilde{\varphi}_\lambda^+$ for $\lambda \in \mathbb{R}_+$, in the following sense.

Proposition 5.6 *For any smooth function f with compact support on $\mathbb{R}$ we have*

$$f(x) = \frac{1}{2\pi}\int_0^\infty \{\langle\varphi_\lambda^-, f\rangle\, \varphi_\lambda^-(x) + \langle\tilde{\varphi}_\lambda^+, f\rangle\, \tilde{\varphi}_\lambda^+(x)\}\frac{\mathrm{d}\lambda}{|a_\lambda|^2},$$

where $\langle\varphi, f\rangle$ denotes $\int_{\mathbb{R}} \overline{\varphi(x)} f(x)\,\mathrm{d}x$.

Thus $\mathcal{H}_0$ is the 'direct integral' of the λ^2-eigenspaces $V_\lambda = V_{-\lambda}$ for $\lambda \in \mathbb{R}_+$, even though V_λ is not contained in $\mathcal{H}$.

In the proposition we can replace φ_λ^- and $\tilde{\varphi}_\lambda^+$ by the 'incoming' basis $\{\varphi_\lambda^+, \tilde{\varphi}_\lambda^-\}$. In fact we have two isomorphisms

$$S_\pm : \mathcal{H}_0 \to L^2(\mathbb{R}_+; \mathbb{C}^2)$$

such that

$$S_+ L_u S_+^{-1} = S_- L_u S_-^{-1}$$

is multiplication by λ^2 on $L^2(\mathbb{R}_+; \mathbb{C}^2)$. The *scattering operator* $S = S_+ S_-^{-1}$ is multiplication by the scattering loop $\{S_\lambda\}$ on $L^2(\mathbb{R}_+; \mathbb{C}^2)$.

We shall return to this description of scattering, in a more general setting, at the end of Section 13.

We shall now give the proof of Proposition 5.6.

Proof If $\mu \in \mathbb{C}$ does not belong to the spectrum of L_u we can define *the resolvent* $R_\mu = (\mu - L_u)^{-1}$. This is a bounded operator, and a holomorphic function of μ in the complex plane cut along the positive real axis, except for poles at $\mu = -\omega_k^2$. We obtain Proposition 5.6 by integrating $R_\mu f$ around a large 'keyhole' contour consisting of the punctured circle

$$C = \{R\mathrm{e}^{\mathrm{i}\theta} : 0 < \theta < 2\pi\}$$

together with paths from R to 0 and 0 to R along the lips of the cut. On the circular part C we have

$$R_\mu - \mu^{-1} = \mu^{-1} R_\mu L_u \,,$$

and $\mu^{-1}R_\mu L_u f$ is $O\,(\mu^{-2})$ as $\mu \to \infty$, so that

$$\int_C R_\mu f \,\mathrm{d}\mu = f \int_C \frac{\mathrm{d}\mu}{\mu} = 2\pi \mathrm{i} f \,.$$

So, by Cauchy's theorem,

$$f = \frac{1}{2\pi \mathrm{i}} \int_0^\infty \check{R}_\mu f d\mu \quad + \sum (\text{ residues of } R_\mu f) \,,$$

where $\check{R}_\mu$ is the jump in R_μ across the real axis.

Now R_μ is an integral operator with kernel

$$R_\mu(x,y) = \begin{cases} \varphi_\lambda^-(x)\tilde{\varphi}_\lambda^+(y)/W_\lambda & \text{if } x \leqslant y \\ \tilde{\varphi}_\lambda^+(x)\varphi_\lambda^-(y)/W_\lambda & \text{if } x \geqslant y, \end{cases}$$

where $\lambda^2 = \mu$, with λ in the upper half-plane, and $W_\lambda = 2\mathrm{i}\lambda a_\lambda$ is the Wronskian $W(\varphi_\lambda^-, \tilde{\varphi}_\lambda^+)$. The poles of R_μ occur at points $\mu_k = \lambda_k^2 = -\omega_k^2$ where $a_\lambda = 0$ and $\tilde{\varphi}_\lambda^+$ and φ_λ^- are proportional. Normalizing the eigenfunction ψ_k appropriately, we can write the residue of $R_\mu f$ at $-\omega_k^2$ as $\langle \psi_k, f \rangle \psi_k$.

Finally, after a little manipulation we find that the jump in R_μ is given by

$$\check{R}_\mu(x,y) = \{\varphi_\lambda^-(x)\varphi_\lambda^-(y)^- \; + \; \tilde{\varphi}_\lambda^+(x)\tilde{\varphi}_\lambda^+(y)^-\}/|a_\lambda|^2 \,.$$

□

The holonomy loop in the periodic case

The role of the loop groups in spectral theory is not a matter of particular boundary conditions. If we are interested in functions u on the circle—thought of as functions on the line satisfying $u(x+2\pi) = u(x)$—then we still have a two-dimensional space V_λ of solutions of $L_u\varphi = \lambda^2\varphi$ for all $\lambda \in \mathbb{C}$. The elements of V_λ are functions on the line, not necessarily periodic, and translation by 2π gives us a map $Q_\lambda : V_\lambda \to V_\lambda$. Identifying V_λ with $\mathbb{C}^2$ by $\varphi \mapsto (\varphi(0), \varphi'(0))$, we have

$Q_\lambda \in SL_2\mathbb{C}$. As $\lambda \to \infty$ the solutions of $L_u\varphi = \lambda^2\varphi$ will resemble $e^{\pm i\lambda x}$, so $Q_\lambda \sim Q_\lambda^\circ$ as $\lambda \to \infty$, where

$$Q_\lambda^\circ = \begin{pmatrix} 1 & 1 \\ 2\pi i & -2\pi i \end{pmatrix} \begin{pmatrix} e^{2\pi i\lambda} & 0 \\ 0 & e^{-2\pi i\lambda} \end{pmatrix}.$$

In this situation the holonomy loop is $\lambda \mapsto (Q_\lambda^\circ)^{-1}Q_\lambda$, which tends to 1 as $\lambda \to \pm\infty$.

6 The non-linear Schrödinger equation and its scattering

In the last section we saw that scattering theory for the operator L_u was slightly complicated by the symmetry $\lambda \leftrightarrow -\lambda$, and the resulting singularity at $\lambda = 0$. The theory is more attractive and geometrical for the closely related operator

$$\Lambda_u = J\frac{d}{dx} + A, \tag{6.1}$$

acting on vector-valued functions $\psi : \mathbb{R} \to \mathbb{C}^2$, where

$$J = \begin{pmatrix} i & 0 \\ 0 & -i \end{pmatrix} \quad \text{and} \quad A = \begin{pmatrix} 0 & -\bar{u} \\ u & 0 \end{pmatrix}$$

for some smooth function $u : \mathbb{R} \to \mathbb{C}$ with compact support, or at least rapidly decreasing at infinity.

The non-linear Schrödinger (NLS) equation is an evolution equation closely analogous to the KdV equation which describes an isospectral deformation of the operator Λ_u. It is

$$i\frac{\partial u}{\partial t} = -\frac{\partial^2 u}{\partial x^2} + 2|u|^2 u,$$

and can be written in the Lax form

$$\frac{d}{dt}\Lambda_u = [P_u, \Lambda_u], \tag{6.2}$$

where

$$P_u = J\left(\frac{d}{dx}\right)^2 + A\frac{d}{dx} + \tfrac{1}{2}(A' - JA^2).$$

The vector space U of rapidly decreasing smooth functions $u : \mathbb{R} \to \mathbb{C}$ has (when regarded as a real vector space) a symplectic bilinear form $\omega : U \times U \to \mathbb{R}$ given by

$$\omega(u_1, u_2) = \operatorname{Im} \int_{\mathbb{R}} u_1 \bar{u}_2 \, \mathrm{d}x \, .$$

The NLS equation is the vector field on the symplectic manifold U generated by the Hamiltonian function $H : U \to \mathbb{R}$, where

$$H(u) = \tfrac{1}{2} \int_{\mathbb{R}} (|u'|^2 + |u|^4) \, \mathrm{d}x \, ,$$

for

$$\begin{aligned} \delta H &= \operatorname{Re} \left\{ \int_{\mathbb{R}} (-u'' + 2|u|^2 u) \delta \bar{u} \, \mathrm{d}x \right\} \\ &= \operatorname{Im} \left\{ \int_{\mathbb{R}} \mathrm{i}(-u'' + 2|u|^2 u) \delta \bar{u} \, \mathrm{d}x \right\} \\ &= \omega(\mathrm{i}(-u'' + 2|u|^2 u), \delta u) \, . \end{aligned}$$

Scattering theory produces a sequence of functions $I_k : U \to \mathbb{R}$ which are in involution for the symplectic structure and are conserved by the NLS flow. Essentially, they are a complete set of 'action' variables for the system. The first three conserved quantities are

$$\begin{aligned} I_0 &= \int |u|^2 \mathrm{d}x \, , \\ I_1 &= \operatorname{Im} \int \bar{u} u' \mathrm{d}x \;=\; \omega(u', u) \, , \\ I_2 &= H. \end{aligned}$$

The flows generated by I_0 and I_1 are multiplication of u by $\mathrm{e}^{\mathrm{i}t}$, and translation of u by t, respectively.

To investigate the scattering for Λ_u let us rewrite the eigenvalue equation $\Lambda_u \psi = \lambda \psi$:

$$\frac{\mathrm{d}\psi}{\mathrm{d}x} = (-\lambda J + JA)\psi = B_\lambda \psi, \quad \text{say.} \tag{6.3}$$

As such, we think of it as describing the parallel transport of the vector-valued function ψ along the line by means of the connection

$$\frac{\mathrm{d}}{\mathrm{d}x} - B_\lambda$$

in the trivial two-dimensional vector bundle on $\mathbb{R}$. If λ is real then B_λ is skew-hermitian with trace zero, i.e. it belongs to the Lie algebra of SU_2.

Solutions of an equation of the form

$$\frac{\mathrm{d}\psi}{\mathrm{d}x} = B(x)\psi \tag{6.4}$$

are best described in terms of a 2×2 *solution matrix* $M(x) = (\psi_1(x)\ \psi_2(x))$, where ψ_1 and ψ_2 are linearly independent solutions of (6.4). The most general solution is then $\psi(x) = M(x)\xi$, where $\xi \in \mathbb{C}^2$ is a constant vector; and for any solution ψ we have

$$\psi(y) = M(y)M(x)^{-1}\psi(x)$$

for any $x, y \in \mathbb{R}$. Any other solution matrix $\widetilde{M}$ for (6.4) is of the form $\widetilde{M} = MC$, where C is a constant invertible matrix.

Applying this to equation (6.3), we have a solution matrix M_λ such that

$$M_\lambda(x) = \begin{cases} \mathrm{e}^{-\lambda Jx} & \text{for } x \ll 0 \\ \mathrm{e}^{-\lambda Jx} g_\lambda & \text{for } x \gg 0, \end{cases}$$

where $g_\lambda \in SU_2$ when $\lambda \in \mathbb{R}$.

As was the case in the preceding section, $g_\lambda \to 1$ rapidly as $\lambda \to \pm\infty$ along $\mathbb{R}$, even for rapidly decreasing u, and $\lambda \mapsto g_\lambda$ extends to a holomorphic map $\mathbb{C} \to SL_2\mathbb{C}$ providing u has compact support. (There is now no problem with $\lambda = 0$.)

The based loop g_λ in SU_2 will be called the *holonomy loop* of the potential $u : \mathbb{R} \to \mathbb{C}$. As SU_2 is three dimensional, while u has only two real components, we cannot expect $u \mapsto g_\lambda$ to be a $1-1$ correspondence. Let us map SU_2 to the Riemann sphere $S^2 = \mathbb{C} \cup \{\infty\}$ by the slightly surprising map

$$\begin{pmatrix} a & -\bar{b} \\ b & \bar{a} \end{pmatrix} \mapsto b/|a| \in \mathbb{C} \cup \{\infty\},$$

and write $\sigma : \mathbb{R} \to S^2$ for the composite

$$\mathbb{R} \to SU_2 \to S^2 .$$

Then σ is a smooth loop in S^2 based at 0, i.e.

$$\sigma(\lambda) \to 0 \ \text{ as } \lambda \to \pm\infty .$$

We shall call σ the *dispersive scattering* of u.

The map $u \mapsto \sigma$ is a kind of non-linear Fourier transform, and it is a local diffeomorphism

$$\{\text{rapidly decreasing functions } u\} \to \{\text{based smooth loops in } S^2\}$$

in the neighbourhood of $u = 0$. In fact its derivative at $u = 0$ is precisely the Fourier transform, as we see from

Proposition 6.5 *For small u we have*

$$g_\lambda = \begin{pmatrix} 1 & -\bar{b}_\lambda \\ b_\lambda & 1 \end{pmatrix}$$

to first order in u, where

$$b_\lambda = \mathrm{i} \int_{\mathbb{R}} u(x) \mathrm{e}^{2\mathrm{i}\lambda x} \mathrm{d}x\,.$$

This follows at once from

Proposition 6.6 *The change δg_λ produced by a small change δu in u is given by*

$$g_\lambda^{-1} \delta g_\lambda = \int_{\mathbb{R}} M_\lambda^{-1} J \delta A M_\lambda \mathrm{d}x\,,$$

where $\delta A = \begin{pmatrix} 0 & -\delta\bar{u} \\ \delta u & 0 \end{pmatrix}$.

Indeed, when $u = 0$ we have $M_\lambda(x) = \mathrm{e}^{-\lambda J x} = \begin{pmatrix} \mathrm{e}^{-\mathrm{i}\lambda x} & 0 \\ 0 & \mathrm{e}^{\mathrm{i}\lambda x} \end{pmatrix}$, so that the integrand in Proposition 6.6 is

$$\begin{pmatrix} 0 & \mathrm{i}\delta\bar{u}\mathrm{e}^{2\mathrm{i}\lambda x} \\ \mathrm{i}\delta u \mathrm{e}^{2\mathrm{i}\lambda x} & 0 \end{pmatrix}.$$

Proposition 6.6, in turn, amounts simply to the fact that if M is a solution matrix of $\psi' = B\psi$ then the first-order change δM in M produced by changing B to $B + b$ is given by

$$\delta M(x) = M(x) \int_{-\infty}^{x} M(y)^{-1} b(y) M(y) \mathrm{d}y\,,$$

as one sees by direct differentiation.

In Section 12 I shall explain how to define a local inverse for the scattering transform $u \mapsto \sigma$. But now I shall describe the global situation in general terms.

On one side we have the vector space U of rapidly decreasing potentials u. It maps to the space ΩS^2 of based loops on the Riemann sphere by the scattering map, which for the moment I shall denote by

$$\Sigma : U \to \Omega S^2 .$$

In a neighbourhood of 0 in U the map Σ is a symplectic diffeomorphism, when ΩS^2 is given its natural symplectic structure coming from that of S^2. (For a proof, see Faddeev and Takhtajan 1987, Chap. III.)

The loop space ΩS^2 has a quite complicated topology. Its fundamental group is $\mathbb{Z}$, generated by a loop in the loop space which 'lassoes' the sphere S^2. For a given $\sigma \in \Omega S^2$ the inverse-image $U_\sigma = \Sigma^{-1}(\sigma)$ consists of the potentials u with a given 'dispersive' component. The loops σ which do not pass through ∞ in S^2 form an open subset Ω' of ΩS^2, and for all $\sigma \in \Omega'$ the spaces U_σ are diffeomorphic to each other, and each is the disjoint union of a sequence of connected components $U_{\sigma,k}$, for $k \geqslant 0$, where $U_{\sigma,k}$ is an algebraic variety of complex dimension $2k$ consisting of potentials with k 'solitons' on top of the dispersive part. In fact, $\Sigma^{-1}(U)$ is isomorphic as a symplectic manifold to $\Omega' \times U_0$. But the map Σ cannot be a fibration, for when σ travels around a generator of the fundamental group of ΩS^2 the 'holonomy' $U_\sigma \to U_\sigma$ increases the soliton number by 1.

7 Families of flat connections and harmonic maps

The non-linear Schrödinger equation has been written in the Lax form $\dot{\Lambda} = [P, \Lambda]$, where

$$\Lambda = J\frac{\mathrm{d}}{\mathrm{d}x} + A .$$

To apply the inverse-scattering method we introduced a solution matrix M_λ satisfying

$$\Lambda M_\lambda = \lambda M_\lambda , \tag{7.1}$$

and we let M_λ evolve in time according to

$$\frac{\partial M_\lambda}{\partial t} = PM_\lambda\,. \tag{7.2}$$

The equation (7.1) can be rewritten as

$$\frac{\partial M_\lambda}{\partial t} = J(A-\lambda)M_\lambda = B_\lambda M_\lambda, \quad \text{say}, \tag{7.3}$$

and this can be substituted in (7.2) to give

$$\frac{\partial M_\lambda}{\partial t} = P_\lambda M_\lambda\,, \tag{7.4}$$

where P_λ is a matrix depending polynomially on λ, but no longer involving $\frac{\mathrm{d}}{\mathrm{d}x}$. The existence of a simultaneous solution M_λ of (7.3) and (7.4) entails

$$\frac{\partial B_\lambda}{\partial t} - \frac{\partial P_\lambda}{\partial x} + [P_\lambda, B_\lambda] = 0 \tag{7.5}$$

for all λ (as $\partial^2 M_\lambda/\partial x\,\partial t = \partial^2 M_\lambda/\partial t\,\partial x$), i.e. that the λ-dependent connection with components

$$\frac{\partial}{\partial x} - B_\lambda\,, \quad \frac{\partial}{\partial t} - P_\lambda, \tag{7.6}$$

in the trivial two-dimensional vector bundle on $\mathbb{R}^2$, is *flat*. The left-hand side of (7.5) is, of course, the curvature of this connection.

In fact the Lax equation $\dot{\Lambda} = [P, \Lambda]$ is exactly equivalent to the flatness of the family of connections (7.6). Surprisingly, perhaps, it seems to be a general principle that any Lax equation can be rewritten as the condition that a 1-parameter family of connections are simultaneously flat. The 'flat connection' formulation is rather more geometrical than the Lax equation, and makes clearer the origin of the conservation laws. For example, if A, and hence B_λ, is periodic in x with period 2π, then the flatness of (7.6) implies that the holonomy of the connection around the rectangle with vertices $(0, t_0)$, $(2\pi, t_0)$, $(2\pi, t_1)$, $(0, t_1)$ is the identity. This means that

$$g_{t_1} = h g_{t_0} h^{-1}\,.$$

where g_t is the holonomy from $(0, t)$ to $(2\pi, t)$, and h is the holonomy from $(0, t_0)$ to $(0, t_1)$.

All the same, it is hard to see any special geometric significance in the 1-parameter family of connections related to the NLS equation. There are other problems where the geometry is much more appealing. The most striking, for me, is the problem of *harmonic maps* from a surface to a compact Lie group G with a bi-invariant Riemannian metric. I shall describe it briefly.

Suppose that $g : \Sigma \to G$ is a smooth map from a surface Σ to G. Then g is harmonic if

$$\mathrm{d}(*g^{-1}\mathrm{d}g) = 0, \tag{7.7}$$

where $g^{-1}\mathrm{d}g$ is a 1-form on Σ with values in the Lie algebra $\mathfrak{g}$ of G, and

$$* : \Omega^1(\Sigma;\mathfrak{g}) \to \Omega^1(\Sigma;\mathfrak{g})$$

is the Hodge $*$-operator which rotates each tangent space to Σ through 90°. For a connection from $A \in \Omega^1(\Sigma;\mathfrak{g})$ we have $A = g^{-1}\mathrm{d}g$ if and only if A is flat, i.e.

$$\mathrm{d}A + \tfrac{1}{2}[A,A] = 0\,,$$

so the harmonic map equation is equivalent to the pair of equations

$$\begin{aligned} \mathrm{d}A + \tfrac{1}{2}[A,A] &= 0 \\ \mathrm{d} * A &= 0\,. \end{aligned} \tag{7.8}$$

These, in turn, are equivalent to the flatness of a 1-parameter family of connections, for the $*$-operator is just one element of a natural action of $\mathbb{C}^\times$ on $\Omega^1(\Sigma;\mathfrak{g})$ which is induced by the complex structure on each tangent space to Σ. If we write, in terms of a local holomorphic parameter z,

$$A = A_1\mathrm{d}z + A_2\mathrm{d}\bar{z}\,,$$

then

$$*A = -\mathrm{i}A_1\mathrm{d}z + \mathrm{i}A_2\mathrm{d}\bar{z}\,,$$

and we can define

$$A_{(\lambda)} = \lambda^{-1}A_1\mathrm{d}z + \lambda A_2\mathrm{d}\bar{z}$$

for any $\lambda \in \mathbb{C}^\times$. The equations (7.8) are equivalent to the flatness of

$$\tfrac{1}{2}(A - A_{(\lambda)}) \quad = \quad \tfrac{1}{2}(1-\lambda^{-1})A_1\mathrm{d}z + \tfrac{1}{2}(1-\lambda)A_2\mathrm{d}\bar{z} \tag{7.9}$$

for all λ, and so we have written the harmonic map equations in the desired form.

We have now left the domain of 'integral systems' in the sense described in Section 2. On the other hand, the reformulation of the harmonic map equations does enable one to solve them completely in many situations. For if the connection (7.9) is flat for all λ then we can write it as $g_\lambda^{-1}\mathrm{d}g_\lambda$ for some map $g_\lambda : \Sigma \to G$. (We should really assume $|\lambda| = 1$ here, to ensure that (7.9) takes values in the real Lie algebra $\mathfrak{g}$.)

Letting λ traverse the circle $|\lambda| = 1$, and observing that we can take $g_1 = 1$, we have thus factorized a harmonic map $g : \Sigma \to G$ canonically as

$$\Sigma \xrightarrow{\{g_\lambda\}} \Omega G \longrightarrow G\,,$$

where ΩG is the space of based loops in G, and $\Omega G \to G$ takes a loop to its value at $\lambda = -1$. But we can say much more. In Section 13 we shall see that the loop space ΩG is naturally a complex manifold in virtue of an isomorphism

$$\Omega G \;\cong\; \mathcal{L}G_{\mathbb{C}}/\mathcal{L}^+G_{\mathbb{C}}\,, \tag{7.10}$$

and the form of the connection (7.9) shows that the map $\Sigma \to \Omega G$ is *holomorphic*. I shall refer to Uhlenbeck (1989) and Segal (1989) for an account of how this holomorphicity can be exploited to describe the harmonic maps from a surface to a group or symmetric space quite explicitly. I should mention, however, that it is at this point that the path followed in those lectures makes contact with the 'twistorial' approach presented by Richard Ward. The loop space ΩG has yet another description as the space of holomorphic $G_{\mathbb{C}}$-bundles on the Riemann sphere S^2 trivialized in a neighbourhood of ∞, so we have arrived at a correspondence between harmonic maps $\Sigma \to G$ and holomorphic bundles on $\Sigma \times X^2$.

8 The KdV equation as an Euler equation

When we consider the operator $L_u = -(\mathrm{d}/\mathrm{d}x)^2 + u$, where u is a smooth function defined, say, in an interval I of $\mathbb{R}$, we are, of course, making use of a chosen parameter x on I. If we change the parameter by a diffeomorphism $x \mapsto \tilde{x}$ then L_u will not transform to an operator of the same form. We can rectify this situation by regarding L_u as an operator

$$L_u : \Omega^{(-1/2)}(I) \to \Omega^{(3/2)}(I), \tag{8.1}$$

where $\Omega^{(a)}(I)$ denotes the a-densities on I, i.e. expressions $\psi(x)(dx)^a$ which transform to $\tilde{\psi}(\tilde{x})(\mathrm{d}\tilde{x})^a$, where

$$\tilde{\psi}(\tilde{x}(x))\left(\frac{\mathrm{d}\tilde{x}}{\mathrm{d}x}\right)^a = \psi(x)\,.$$

Transforming L_u as indicated by (8.1) changes L_u to $L_{\tilde{u}}$, where

$$\tilde{u}(\tilde{x}(x))\left(\frac{\mathrm{d}\tilde{x}}{\mathrm{d}x}\right)^2 + S(\tilde{x},x) = u(x), \tag{8.2}$$

and $S(\tilde{x},x)$ is the *Schwarzian derivative*

$$S(\tilde{x},x) = \frac{1}{2}\frac{\tilde{x}'''}{\tilde{x}'} - \frac{3}{4}\left(\frac{\tilde{x}''}{\tilde{x}'}\right)^2.$$

It would take us too far afield to explain fully the reasons for treating L_u as just described (see Segal 1991). But an important fact is that to give the operator L_u is exactly equivalent to giving a *projective coordinate* on I, i.e. an immersion $f_u : I \to \mathbb{P}^1_{\mathbb{R}}$ of I in the real projective line, up to a projective transformation of $\mathbb{P}^1_{\mathbb{R}}$. (Given L_u, we define $f_u(x) = \psi_1(x)/\psi_0(x) \in \mathbb{R}\cup\{\infty\}$, where ψ_0 and ψ_1 are independent solutions of $L_u\psi = 0$.) Changing L_u to $L_{\tilde{u}}$ by (8.2) corresponds simply to

$$f_{\tilde{u}}(\tilde{x}(x)) = f_u(x).$$

For an infinitesimal movement given by a vector field $\xi = \xi(x)\mathrm{d}/\mathrm{d}x$ on I the change in u corresponding to (8.2) is

$$D_\xi u = \xi u' + 2\xi' u + \tfrac{1}{2}\xi'''. \tag{8.3}$$

In this notation the KdV equation is

$$\frac{\partial u}{\partial t} = 2D_u u. \tag{8.4}$$

We have still to explain in what sense equation (8.4) is an Euler equation. If we omitted the term $S(\tilde{x},x)$ then (8.2) would say that u transforms under diffeomorphisms as an element of $\Omega^{(2)}(I)$. The affine transformation (8.2) of the operators L_u is simplest treated as a linear transformation of the vector space $\mathcal{O}$ of all operators of the form $-k(\mathrm{d}/\mathrm{d}x)^2 + u$, with k constant, on which diffeomorphisms act by $(k, u) \mapsto (k, \tilde{u})$ where

$$\tilde{u}(\tilde{x}(x)) \left(\frac{\mathrm{d}\tilde{x}}{\mathrm{d}x}\right)^2 + kS(\tilde{x}, x) = u(x)\,.$$

Thus $\mathcal{O}$ fits into an exact sequence of representations of the group $\mathrm{Diff}(I)$ of diffeomorphisms of I:

$$0 \to \Omega^{(2)}(I) \to \mathcal{O} \to \mathbb{R} \to 0\,. \tag{8.5}$$

The quadratic densities $\Omega^{(2)}(I)$ are naturally in duality with $\mathrm{Vect}_{\mathrm{cpt}}(I)$, the Lie algebra of vector fields with compact support on I, for the pointwise product of a vector field and a quadratic density is a 1-form, which can naturally be integrated over I. The sequence (8.5) is therefore in duality with a sequence

$$0 \to \mathbb{R} \to \mathrm{Vect}^{\sim}_{\mathrm{cpt}}(I) \to \mathrm{Vect}_{\mathrm{cpt}}(I) \to 0 \tag{8.6}$$

of representations of $\mathrm{Diff}(I)$. The action of the Lie algebra $\mathrm{Vect}(I)$ of $\mathrm{Diff}(I)$ on $\mathrm{Vect}^{\sim}_{\mathrm{cpt}}(I)$ is easily seen to induce a Lie algebra structure on $\mathrm{Vect}^{\sim}_{\mathrm{cpt}}(I)$, with the image of $\mathbb{R}$ in its centre, so that (8.6) is an exact sequence of Lie algebras.

It is true, but not so obvious (Segal 1981), that $\mathrm{Vect}^{\sim}_{\mathrm{cpt}}(I)$ is the Lie algebra of an infinite dimensional Lie group G which is a central extension of $\mathrm{Diff}_{\mathrm{cpt}}(I)$ by $\mathbb{R}$. The KdV equation is an Euler equation for G of the standard form described in Section 3. Recall that to get an Euler equation we choose a (non-invariant) quadratic form q on the dual of the Lie algebra of G. In the present case we take

$$q(k, u) = \int_I u^2 \mathrm{d}x\,.$$

The diffeomorphism invariance is broken here by the choice of the parameter x on I. (Unfortunately q is defined only on the subspace of $\mathcal{O}$ where u has compact support.) As a map $\mathcal{O}_{\mathrm{cpt}} \to \mathrm{Vect}^{\sim}_{\mathrm{cpt}}(I)$, the form q takes L_u to $2u\frac{\mathrm{d}}{\mathrm{d}x}$, and so the resulting Euler equation (3.2) is the KdV equation (8.4).

To obtain a family of commuting conserved quantities for the equation the method proposed in Section 3 was to take a G-invariant function $F : \mathcal{O}_{\mathrm{cpt}} \to \mathbb{R}$, and then to define

$$F_\lambda : \mathcal{O}_{\mathrm{cpt}} \to \mathbb{R}$$

by $F_\lambda(L_u) = F(L_u - \lambda\beta)$ for some fixed β. It is simplest to understand the situation in the case of *periodic* functors u, in order to avoid questions of boundary conditions. Then the only obvious invariant function of an operator $L_u \in \mathcal{O}$ is the trace of its holonomy matrix g. (If V is the two-dimensional space of solutions of $L_u\varphi = 0$ on the line, then $g : V \to V$ is the operation of translating φ by one period.) The obvious choice for β is the constant 1 (i.e., strictly speaking, $\beta = (\mathrm{d}\theta)^2$). Then

$$F_\lambda(L_u) = \operatorname{trace}(g_\lambda)\,,$$

where g_λ is the holonomy loop of L_u. These are the conserved quantities of the KdV flow which we have already identified, and we now know that they are in involution.

9 Determinants and holonomy

In Section 1 I mentioned that the transmission coefficient T_λ of the operator $L_u = -(\mathrm{d}/\mathrm{d}x)^2 + u$ was 'essentially' the characteristic polynomial $\det(\lambda^2 - L_u)$. A similar result holds for the operator

$$\Lambda_u = J\frac{\mathrm{d}}{\mathrm{d}x} + A \tag{9.1}$$

discussed in Section 6. This section is devoted to elucidating these assertions, and to the relation between such infinite dimensional determinants and holonomy in general.

The clearest case to treat is that of operators on the circle, for then the spectrum of L_u or Λ_u is *discrete*, as is the case for any elliptic operator on a compact manifold. If u is periodic with period 2π then the two-dimensional space V_λ of solutions (on the line) of $L_u\varphi = \lambda^2\varphi$, or of $\Lambda_u\psi = \lambda\psi$, possesses a linear automorphism g_λ induced by translating φ or ψ by 2π. This is the *holonomy* of L_u or Λ_u. A solution of $g_\lambda\varphi = \varphi$ or $g_\lambda\psi = \psi$ is the same thing as a periodic eigenfunction of L_u or Λ_u, so $\det(g_\lambda - 1)$ vanishes precisely when λ is an eigenvalue of the operator on the circle. That makes the following result plausible.

Proposition 9.2 *We have*

$$\det(\lambda^2 - L_u) = \det(g_\lambda - 1)$$

and also

$$\det(\lambda - \Lambda_u) = \det(g_\lambda - 1)\,.$$

Remark As $\det(g_\lambda) = 1$ we have $\det(g_\lambda - 1) = 2 -$ trace (g_λ).

To make sense of Proposition 9.2 we must explain how to define the infinite dimensional determinants. Let us begin very heuristically, with the case $u = 0$. Then the eigenfunctions of $L_u = -(\mathrm{d}/\mathrm{d}x)^2$ are $\mathrm{e}^{\mathrm{i}nx}$ for $n \in \mathbb{Z}$, and the eigenvalues are n^2, each except 0 being of multiplicity two. The holonomy is

$$g_\lambda = \begin{pmatrix} \mathrm{e}^{2\pi\mathrm{i}\lambda} & 0 \\ 0 & \mathrm{e}^{-2\pi\mathrm{i}\lambda} \end{pmatrix},$$

so that

$$\begin{aligned} \det(g_\lambda - 1) &= (\mathrm{e}^{2\pi\mathrm{i}\lambda} - 1)(\mathrm{e}^{-2\pi\mathrm{i}\lambda} - 1) \\ &= 4\sin^2 \pi\lambda\,. \end{aligned}$$

Formally we have

$$\begin{aligned} \det(\lambda^2 - L_u) &= \lambda^2 \prod_{n>0} (\lambda^2 - n^2)^2 \\ &= \prod_{n>0} n^4 \cdot \left\{ \lambda \prod \left(1 - \frac{\lambda^2}{n^2}\right) \right\}^2 \\ &= \prod_{n>0} n^4 \cdot \frac{\sin^2 \pi\lambda}{\pi^2}, \end{aligned}$$

in view of the product expansion of $\sin \pi\lambda$.

Similarly, the eigenvalues of Λ_u are the positive and negative integers, with multiplicity two, while g_λ is the same as for L_u. Thus

$$\det(\lambda - \Lambda_u) = \prod_{n\in\mathbb{Z}} (\lambda - n)^2\,,$$

which is formally the same as $\det(\lambda^2 - L_u)$.

The divergent infinite products occurring here can be regularized by the ζ-function method. If an operator T has eigenvalues $\{\lambda_n\}$ we define

$$\zeta_T(s) = \text{ trace } (T^{-s}) = \Sigma\lambda_n^{-s}$$

for all $s \in \mathbb{C}$ with sufficiently large real part. For the operators T we are concerned with, the function ζ_T can be continued analytically to a function which is meromorphic in the whole complex plane, and we can define

$$\det_\zeta(T) = e^{-\zeta'_T(0)},$$

motivated by

$$\zeta'_T(s) = -\sum \frac{\log \lambda_n}{\lambda_n^s}.$$

This procedure is very familiar when the operator T is positive, but it works in any case, providing we choose a cut from 0 to ∞ in the complex plane which does not pass through any λ_n, so that λ_n^{-s} can be defined as a holomorphic function of s.

Example If $\lambda_n = n - a$, where $a \notin \mathbb{Z}$, then

$$\begin{aligned}\zeta_T(s) &= \sum_{n\in\mathbb{Z}} \frac{1}{(n-a)^s} \\ &= \int_\gamma \frac{(z-a)^{-s}}{\mathrm{e}^{2\pi \mathrm{i} z} - 1}\,\mathrm{d}z,\end{aligned}$$

where, in the complex plane cut from a to $a - \mathrm{i}\infty$, the contour γ goes from $a - \mathrm{i}\infty$ to a on one side of the cut, and back again on the other. This formula is true because when $\mathrm{Re}(s)$ is large one can complete γ to a closed contour by adding a very large circle with centre a and then can calculate the integral by means of residues, noting that $(\mathrm{e}^{2\pi \mathrm{i} z} - 1)^{-1}$ has a pole of residue $1/2\pi\mathrm{i}$ at each integer. The integral formula, however, defines an entire function of s. By differentiating under the integral sign we get

$$\begin{aligned}-\zeta'{}_T(0) &= \int_\gamma \frac{\log(z-a)}{\mathrm{e}^{2\pi \mathrm{i} z} - 1}\,\mathrm{d}z \\ &= -2\pi\mathrm{i}\int_0^\infty \frac{-i\,\mathrm{d}t}{\mathrm{e}^{2\pi\mathrm{i}(a-\mathrm{i}t)} - 1} \\ &= -\log(1 - \mathrm{e}^{2\pi\mathrm{i}a}),\end{aligned}$$

corresponding to the formula

$$\prod\nolimits_{n\in\mathbb{Z}}^{(\zeta)} (n-a) = 1 - \mathrm{e}^{2\pi \mathrm{i} a}. \tag{9.3}$$

The method of ζ-function regularization has the property that if $\{\lambda_n\}$ and $\{\mu_n\}$ are sequences such that $\prod(\mu_n/\lambda_n)$ converges, then

$$\prod^{(\zeta)}\lambda_n \cdot \prod(\mu_n/\lambda_n) = \prod^{(\zeta)}\mu_n\,,$$

where $\prod^{(\zeta)}$ denotes the regularized product. That justifies our heuristic calculations above, using

$$\prod\nolimits^{(\zeta)}_{n\neq 0} n = 2\pi\mathrm{i}\,,$$

which can be obtained by letting $a \to 0$ in (9.3).

Applied to an operator T which undergoes a small change δT this gives us

Proposition 9.4 *If $T^{-1}\delta T$ is of trace class, then*

$$\delta \log \det\nolimits_{(\zeta)}(T) = \ \mathrm{trace}\ (T^{-1}\delta T)\,.$$

We can use this to prove Proposition 9.2 by showing that both sides have the same derivative with respect to u, for we have already treated the case $u = 0$. Taking $T = \lambda^2 - L_u$, and $\delta T = -\delta u$, we have

$$\delta \log \det\nolimits_{(\zeta)}(\lambda^2 - L_u) = -\mathrm{trace}\ \{(\lambda^2 - L_u)^{-1}\delta u\}\,.$$

Now $(\lambda^2 - L_u)^{-1}$, which is called the *resolvent* of L_u, is an integral operator defined by a kernel $G_\lambda(x, y)$ which is continuous in both variables. It is therefore of trace class, and

$$\delta \log \det\nolimits_{(\zeta)}(\lambda^2 - L_u) = -\int_0^{2\pi} G_\lambda(x, x)\delta u(x)\mathrm{d}x. \qquad (9.5)$$

For Λ_u the position is slightly more complicated. The operator $(\lambda - \Lambda_u)^{-1}$ is given by a (2×2)-matrix valued kernel, which I shall again denote by $G_\lambda(x, y)$. It is a solution of

$$\left(J\frac{\mathrm{d}}{\mathrm{d}x} + A\right) G_\lambda(x, y) = \delta(x - y)1\,,$$

and so has a jump of $-J$ on the diagonal. This means that the resolvent is not quite of trace class, though the formula

$$\delta \log \det\nolimits_{(\zeta)}(\lambda - \Lambda_u) = -\int_0^{2\pi} \mathrm{tr}\{G_\lambda(x, x)\delta A(x)\}\,\mathrm{d}x\,. \qquad (9.6)$$

where $G_\lambda(x,x)$ denotes either of the two values, and tr denotes the (2×2)-matrix trace, in fact gives the right answer.[6] (The point here is simply the difference between the convergent product

$$\prod_{n>0}\left(1-\frac{\lambda^2}{n^2}\right)$$

and the equal but slightly less convergent product

$$\prod_{n\neq 0}\left(1-\frac{\lambda}{n}\right),$$

and I shall ignore it.)

To calculate (9.6) we need

Lemma 9.7 *The kernel of the resolvent of Λ_u is given by*

$$G_\lambda(x,y) = M_\lambda(x)(g_\lambda - 1)^{-1}M_\lambda(y)^{-1}J\,,$$

if $y < x < y + 2\pi$, where M_λ is a solution matrix of $\Lambda_u M_\lambda = \lambda M_\lambda$.

Proof As a function of x for fixed y, $G_\lambda(x,y)$ must satisfy $\Lambda_u G_\lambda = \lambda G_\lambda$ when $x \neq y$, it must be periodic in x with period 2π, and it must jump by $-J$ when $x = y$. So

$$G(x,y) = M(x)C\,,$$

for $y < x < y + 2\pi$, where $M(y+2\pi)C + (-J) = M(y)C$. But $M(y+2\pi) = M(y)g_\lambda$, and this gives us Lemma 9.7. □

We now have

$$\begin{aligned}
&\operatorname{tr}\{(\lambda - L_u)^{-1}\delta A\}\\
&= \int_0^{2\pi} \operatorname{tr}\{M_\lambda(x)(g_\lambda - 1)^{-1}M_\lambda(x)^{-1}J\delta A(x)\}\,\mathrm{d}x\\
&= \operatorname{tr}\left\{(g_\lambda - 1)^{-1}\int M_\lambda(x)^{-1}J\delta A(x)M_\lambda(x)\,\mathrm{d}x\right\}\\
&= \operatorname{tr}\{(g_\lambda - 1)^{-1}g_\lambda^{-1}\delta g_\lambda\}\\
&= \delta \log\det(g_\lambda^{-1} - 1)\\
&= \delta \log\det(g_\lambda - 1)\,,
\end{aligned}$$

[6]Notice that $\operatorname{tr}(J\delta A)$ is identically zero.

where we have used the formula in Proposition 6.6 for $g_\lambda^{-1}\delta g_\lambda$, and, in the last line, the fact that $\det(-g_\lambda) = 1$.

That completes the proof of Proposition 9.2 for Λ_u. The case of L_u is easier.

Let us now look briefly at operators on the line, rather than the circle. Then the spectrum of L_u is not discrete, and $(\lambda^2 - L_u)^{-s}$ is never of trace class, so we cannot define the determinant by means of a ζ-function. On the other hand, as u varies in the space U of rapidly decreasing functions on the line the operator $(\lambda^2 - L_u)^{-1}\delta L_u$ is of trace class. So

$$\operatorname{trace}\{(\lambda^2 - L_u)^{-1}\delta L_u\}$$

is a 1-form on U, and is easily seen to be closed. It can be regarded as

$$\mathrm{d}\log\det(\lambda^2 - L_u)\,,$$

which is thereby defined up to a factor independent of u.

Proposition 9.8 *We have*

$$\det(\lambda^2 - L_u) = a_\lambda$$

up to a factor independent of u. Here λ is chosen in the upper half-plane, and a_λ is the leading entry of the holonomy matrix g_λ.

Proof The coefficient a_λ is defined by means of the solution ϕ_λ^- of $L_u\phi = \lambda^2\phi$ which becomes $\mathrm{e}^{-\mathrm{i}\lambda x}$ as $x \to -\infty$. In fact

$$a_\lambda = \lim_{x\to\infty} \phi_\lambda^-(x)\mathrm{e}^{\mathrm{i}\lambda x}\,.$$

The usual argument tells us that if u changes by δu then ϕ_λ^- changes by $\delta\phi_\lambda^-$, where

$$\delta\phi_\lambda^-(x) = \int_{-\infty}^{x} \widetilde{G}_\lambda(x,y)\delta u(y)\phi_\lambda^-(y)\,\mathrm{d}y\,,$$

and $\widetilde{G}_\lambda$ is the Green's function of $\lambda^2 - L_u$ which vanishes when $x < y$. We have

$$\widetilde{G}_\lambda(x,y) = \{\phi_\lambda^-(x)\widetilde{\phi}_\lambda^+(y) - \widetilde{\phi}_\lambda^+(x)\phi_\lambda^-(y)\}/W(\phi_\lambda^-, \widetilde{\phi}_\lambda^+)$$

when $x \geqslant y$. Then

$$\phi_\lambda^-(x)^{-1}\widetilde{G}_\lambda(x,y) \quad \longrightarrow \quad \frac{1}{2\mathrm{i}\lambda a_\lambda}\widetilde{\phi}_\lambda^+(y)$$

as $x \to \infty$, so

$$\begin{aligned} a_\lambda^{-1}\delta a_\lambda &= \lim \phi_\lambda^-(x)^{-1}\delta\phi_\lambda^-(x) \\ &= \int_{\mathbb{R}} \widetilde{\phi}_\lambda^+(y)\delta u(y)\phi_\lambda^-(y)\frac{\mathrm{d}y}{2\mathrm{i}\lambda a_\lambda} \\ &= \int_{\mathbb{R}} G_\lambda(y,y)\delta u(y)\,\mathrm{d}y\,, \end{aligned}$$

where G_λ is the kernel of the integral operator $(\lambda^2 - L_u)^{-1}$ acting on $L^2(\mathbb{R})$. Thus

$$a_\lambda^{-1}\delta a_\lambda = \operatorname{trace}\{(\lambda^2 - L_u)^{-1}\delta u\}\,,$$

as we want. □

10 Local conservation laws

For the operator L_u on the line we saw in Section 1 that for each λ the transmission coefficient T_λ is conserved by the KdV flows. As $\lambda \to \infty$ we know that $T_\lambda \to 1$, and we even have an asymptotic expansion

$$\log T_\lambda \sim \frac{I_0}{\lambda} + \frac{I_1}{\lambda^2} + \cdots . \tag{10.1}$$

I shall now discuss why the coefficients I_k are *local* functionals of u in the sense of Section 1. This would not be true, for instance, for the coefficients of the expansion of T_λ itself.

Asymptotic expansions can be integrated, so it is enough to show that the expansion of $\mathrm{d}/\mathrm{d}\lambda \log T_\lambda$ is local. I shall begin with some rather general remarks. In §9 we saw that $T_\lambda^{-1} = a_\lambda$ is in some sense the determinant of $\lambda^2 - L_u$, and we should have

$$\frac{\mathrm{d}}{\mathrm{d}\lambda}\log\det(\lambda^2 - L_u) = 2\lambda \operatorname{trace}\,(\lambda^2 - L_u)^{-1}\,.$$

The resolvent $(\lambda^2 - L_u)^{-1}$ is an integral operator with kernel $G_\lambda(x, y)$. The result we want is the existence of an asymptotic expansion

$$G_\lambda(x,x) \sim \sum_{k\geqslant 0} \frac{F_k(x)}{\lambda^{k+1}}, \tag{10.2}$$

where F_k are polynomials in $u(x), u'(x), \ldots$. This is a particular case of one of the central facts in the analysis of elliptic operators. It is perhaps more familiar as the 'asymptotic expansion of the heat kernel'. The *heat kernel* of L_u is the kernel $K_t(x, y)$ of the integral operator e^{-tL_u}. As $t \to 0$ we have

$$K_t(x,x) \sim \sum_{k\geqslant 0} F_k(x) t^{(k-1)/2}, \tag{10.3}$$

and 10.2 is simply the Laplace transform of this:

$$(\lambda^2 - L_u)^{-1} = -\int_0^\infty \mathrm{e}^{t(\lambda^2 - L_u)}\,\mathrm{d}t\,,$$

so that

$$G_\lambda(x,x) = \int_0^\infty K_t(x,x)\mathrm{e}^{t\lambda^2}\,\mathrm{d}t\,.$$

(Of course we need $\mathrm{Re}(\lambda^2) < 0$ to make these integrals converge.)

Rather than pursuing this too schematic discussion I shall sketch a direct argument for the localness. One well-known place in analysis where an asymptotic expansion has local coefficients is the following. If f is a smooth function with compact support on the line, and

$$F(\lambda) = \int_0^\infty f(x)\mathrm{e}^{-\lambda x}\,\mathrm{d}x\,,$$

then by integrating by parts repeatedly we find

$$F(\lambda) \sim \frac{f(0)}{\lambda} + \frac{f(0)}{\lambda^2} + \frac{f''(0)}{\lambda^3} + \cdots \tag{10.4}$$

as $\lambda \to \infty$. Thus although $F(\lambda)$ depends on the whole function f its expansion at ∞ depends only on the values of f and its derivatives at $x = 0$.

The heuristic argument above suggests that

$$a_\lambda^{-1}\,\frac{\mathrm{d}}{\mathrm{d}\lambda}\,a_\lambda = \int_{\mathbb{R}} G_\lambda(x,x)\,\mathrm{d}x\,,$$

but that cannot be correct, as

$$G_\lambda(x,x) = (2\mathrm{i}\lambda)^{-1} + O(\lambda^{-2})\,,$$

and the integral is divergent. The calculation of $a_\lambda^{-1}\,\delta/\delta u\,a_\lambda$ at the end of Section 9 can, however, be adapted to give

Lemma 10.5 $a_\lambda^{-1}\,\dfrac{\mathrm{d}}{\mathrm{d}\lambda}\,a_\lambda = \displaystyle\int_{\mathbb{R}} \left\{ G_\lambda(x,x) - \frac{1}{2\mathrm{i}\lambda} \right\} \mathrm{d}x\,.$

Our task, therefore, is to justify (10.2), as the heuristic argument predicted. Thinking of the resolvent $R_\lambda = (\lambda^2 - L_u)^{-1}$ as a perturbation of $R_\lambda^0 = (\lambda^2 - L_0)^{-1}$, we have a convergent expansion

$$R_\lambda = R_\lambda^0 + R_\lambda^0 u R_\lambda^0 + R_\lambda^0 u R_\lambda^0 u R_\lambda^0 + \cdots\,. \qquad (10.6)$$

The kernel of R_λ^0 is

$$G_\lambda^0(x,y) = \frac{1}{2\mathrm{i}\lambda}\mathrm{e}^{\mathrm{i}\lambda|x-y|}\,,$$

so the contribution to $G_\lambda(x,x)$ of the second term in (10.6) is

$$\begin{aligned}\int_{\mathbb{R}} G_\lambda^0(x,y)u(y)G_\lambda^0(y,x)\,\mathrm{d}y \;&=\; \frac{1}{(2\mathrm{i}\lambda)^2}\int u(y)\mathrm{e}^{2\mathrm{i}\lambda|x,y|}\,\mathrm{d}y \\ &\sim\; 2\sum_{k\geqslant 1}\left(\frac{-1}{2\mathrm{i}\lambda}\right)^{2k+3} u^{2k}(x)\end{aligned}$$

by 10.4. This is an asymptotic expansion of the type we want.

The general term in (10.6) contributes

$$\frac{1}{(2\mathrm{i}\lambda)^{k+1}}\int_0^\infty \mathrm{e}^{\mathrm{i}\lambda t}U_k(x,t)\,\mathrm{d}t\,,$$

where $U_k(x,t)$ is the integral of

$$u(y_1)u(y_2)\ldots u(y_k)$$

over the compact polygon $P(x,t)$ in $\mathbb{R}^k$ defined by

$$|x - y_1| + |y_1 - y_2| + \cdots + |y_k - x| = t\,.$$

Expanding each $u(y_i)$ in a Taylor series at $y_i = x$ it is clear that we have

$$U_k(x,t) = \sum_{m>0} U_{km}(x)t^m\,,$$

where U_{km} is a polynomial of degree k in $u(x), u'(x), \ldots$. Applying 10.4 again gives us (10.2.

11 The classical moment problem

The analytic features of inverse-scattering theory are very much the same as those that arise in the solution of a much simpler—purely linear—classical problem (Akhiezer 1965), that of determining a probability measure μ on the line when its *moments*

$$m_n = \int_{\mathbb{R}} \lambda^n \mathrm{d}\mu(\lambda)$$

are given for all $n \geqslant 0$. In this section I shall discuss the solution of the moment problem, though it has no direct relation to integrable systems, in the hope that it will illuminate the treatment of inverse-scattering.

Let us first suppose that μ is supported in the compact interval $[-R, R]$, and is given by a smooth positive density function

$$\mathrm{d}\mu(\lambda) = u(\lambda)\mathrm{d}\lambda\,.$$

The sequence $\{m_n\}$ then satisfies the growth condition

$$|m_n| \leqslant CR^n$$

for some C, and also positivity conditions reflecting the positivity of μ. A measure with compact support on $\mathbb{R}$ is the same thing as a continuous linear map

$$C(\mathbb{R}) \to \mathbb{R}\,,$$

where $C(\mathbb{R})$ is the vector space of continuous functions on $\mathbb{R}$ with the topology of uniform convergence on compact sets, and so the problem can be formulated as that of extending a given linear map

$$\mathbb{R}[\lambda] \to \mathbb{R}$$

on the polynomial ring in the indeterminate λ (taking λ^n to m_n) to the vector space $C(\mathbb{R})$, in which $\mathbb{R}[\lambda]$ is a dense subspace. One can prove the extendibility abstractly in that context without providing any way of recovering the density u from the moments $\{m_n\}$.

One explicit way of actually finding u, in principle, is to consider the function F of a complex variable λ defined by

$$F(\lambda) = \int_{\mathbb{R}} \frac{\mathrm{d}\mu(t)}{\lambda - t} = \frac{m_0}{\lambda} + \frac{m_1}{\lambda^2} + \frac{m_2}{\lambda^3} + \cdots . \qquad (11.1)$$

The integral expression for F shows that it is defined and holomorphic in the complement of the segment $[-R; R]$ of the real axis. The series expansion, however, converges only for $|\lambda| > R$. If μ is given by a smooth density u then we can recover u from F as its 'jump' across the cut $[-R, R]$ in the real axis:

$$u(\lambda) = \lim_{\varepsilon \to 0} \frac{1}{2\pi \mathrm{i}} \{F(\lambda - \mathrm{i}\varepsilon) - F(\lambda + \mathrm{i}\varepsilon)\} .$$

This relation holds because the function of t

$$\frac{1}{2\pi \mathrm{i}} \left\{ \frac{1}{\lambda - t - \mathrm{i}\varepsilon} - \frac{1}{\lambda - t + \mathrm{i}\varepsilon} \right\}$$

tends to the delta-function at $t = \lambda$ as $\varepsilon \to 0$. The procedure—not very practical—for finding u from the $\{m_n\}$ is to define a holomorphic function in $|\lambda| > R$ by the series $\Sigma m_n \lambda^{-n-1}$, continue it analytically to the complement of the cut, and then measure the jump. The crucial thing is that

(i) there is a unique function F holomorphic in the complement of the cut which has a given jump across the cut and vanishes at ∞, and

(ii) F is completely determined by its series expansion at $\lambda = \infty$.

Example If μ is the uniform distribution $\mathrm{d}\lambda$ on $[-1, 1]$ then odd moments vanish, while

$$m_n = \frac{2}{n+1} \quad \text{if } n \text{ is even.}$$

We have

$$F(\lambda) = \log \frac{\lambda + 1}{\lambda - 1} = \log \left| \frac{\lambda + 1}{\lambda - 1} \right| + \mathrm{i}\theta ,$$

where $\theta \in (-\pi, \pi)$ is the angle subtended by $\{-1, 1\}$ at λ, which jumps by 2π when one crosses $[-1, 1]$.

The situation is more interesting if we consider a measure μ which decays rapidly at infinity but does not have compact support. Then 11.1 is an asymptotic expansion which does not converge anywhere.

Example If μ is $e^{-\frac{1}{2}\lambda^2}\,d\lambda$ then odd moments vanish, and

$$m_{2n} = \sqrt{2\pi}(2n-1)(2n-3)\cdots 3.1\,.$$

In this situation the integral in (11.1) defines a holomorphic function F_+ in the upper half-plane, and another holomorphic function F_- in the lower half-plane. Both F_+ and F_- tend to zero as $\lambda \to \infty$, and both have the *same* asymptotic expansion $\Sigma m_n \lambda^{-n-1}$ as $\lambda \to \infty$. Furthermore, the pair of holomorphic functions (F_+, F_-) is determined by the jump $F_+ - F_-$ on the real axis (providing $F_\pm \to 0$ as $\lambda \to \infty$). However $F_\pm$ are not necessarily determined by the asymptotic expansion at ∞.

Besides the integral formula (11.1) the functions $F_\pm$ can be calculated from u by Fourier transformation. For a rapidly decreasing function f on the line is the boundary value of a bounded holomorphic function in the upper half-plane if and only if the Fourier transform

$$\hat{f}(p) = \frac{1}{2\pi}\int_{\mathbb{R}} f(x)e^{-ipx}\,dx$$

vanishes for $p \leqslant 0$. This means that we have

$$u(\lambda) = F_+(\lambda) - F_-(\lambda)$$

where

$$F_+(\lambda) = \int_0^\infty e^{ip\lambda}\hat{u}(p)\,dp$$

and

$$F_-(\lambda) = -\int_{-\infty}^0 e^{ip\lambda}\hat{u}(p)dp\,.$$

As I said at the beginning, the moment problem is related to inverse scattering only by analogy. The function F above is analogous to the eigenfunction φ_λ of L_u, while the measure μ is analogous to the reflection coefficient.

12 Inverse scattering

In Section 5 we saw that there are solutions φ_λ^- and $\tilde{\varphi}_\lambda^+$ of the equation $L_u\varphi = \lambda^2\varphi$ which are defined and holomorphic for λ in the upper half-plane, and such that, at any point x of the line,

$$\varphi_\lambda^-(x)\mathrm{e}^{\mathrm{i}\lambda x} \to 1 \quad \text{and} \quad \tilde{\varphi}_\lambda^+(x)\mathrm{e}^{-\mathrm{i}\lambda x} \to 1$$

as $\lambda \to \infty$. The corresponding statement for the equation

$$\Lambda_u\psi = \lambda\psi, \tag{12.1}$$

where

$$\Lambda_u = J\frac{\mathrm{d}}{\mathrm{d}x} + A$$

as in Section 6, is that there is a solution matrix M_λ^+ of (12.1) defined and holomorphic for λ in the upper half-plane, and such that, for each x,

$$\chi_\lambda^+(x) = M_\lambda^+(x)\mathrm{e}^{\lambda Jx}$$

tends to 1 as $\lambda \to \infty$. In fact we can take M_λ^+ to be the matrix whose columns are $\psi_\lambda^{(1)}, \widetilde{\psi}_\lambda^{(2)}$, where

$$\psi_\lambda^{(1)}(x) \sim \begin{pmatrix} \mathrm{e}^{-\mathrm{i}\lambda x} \\ 0 \end{pmatrix} \qquad \text{for } x \ll 0,$$

$$\widetilde{\psi}_\lambda^{(2)}(x) \sim \begin{pmatrix} 0 \\ \mathrm{e}^{\mathrm{i}\lambda x} \end{pmatrix} \qquad \text{for } x \gg 0.$$

Similarly, there is a solution matrix $\widetilde{M}_\lambda^-$ holomorphic for λ in the lower half-plane, with columns $\widetilde{\psi}_\lambda^{(1)}, \psi_\lambda^{(2)}$, such that

$$\widetilde{\chi}_\lambda^-(x) = \widetilde{M}_\lambda^-(x)\mathrm{e}^{\lambda Jx} \to 1 \quad \text{as } \lambda \to \infty\,.$$

There is some advantage, however, in replacing $\widetilde{M}_\lambda^-$ by $M_\lambda^- = a_\lambda^{-1}\widetilde{M}_\lambda^-$, where $a_\lambda = \det(M_\lambda^+)$. When this is done M_λ^- will only be *meromorphic* for λ in the lower half-plane, but we shall have the unitarity property

$$(M_{\bar{\lambda}}^-)^* = (M_\lambda^+)^{-1}. \tag{12.2}$$

If λ is real then both M_λ^+ and M_λ^- are defined, and are related by

$$M_\lambda^+(x) = M_\lambda^-(x)S_\lambda\,,$$

where

$$S_\lambda = \frac{1}{|a_\lambda|^2}\begin{pmatrix} 1 & b_\lambda \\ -c_\lambda & 1 \end{pmatrix}$$

is the scattering loop obtained from the holonomy loop[7]

$$g_\lambda = \begin{pmatrix} a_\lambda & b_\lambda \\ c_\lambda & d_\lambda \end{pmatrix}$$

in SU_2, which is defined by

$$(\widetilde{\psi}_\lambda^{(1)}, \widetilde{\psi}_\lambda^{(2)})g_\lambda = (\psi_\lambda^{(1)}\psi_\lambda^{(2)}).$$

We have

$$\chi_\lambda^+(x) = \chi_\lambda^-(x)\cdot \mathrm{e}^{-\lambda Jx}S_\lambda \mathrm{e}^{\lambda Jx}. \tag{12.3}$$

This equation is the basis of inverse scattering theory. For each x we have a based loop $\lambda \mapsto \mathrm{e}^{-\lambda Jx}S_\lambda \mathrm{e}^{\lambda Jx}$ in $GL_2\mathbb{C}$, which is known if the scattering loop S_λ is known. We want to solve (12.3) for the matrix-valued functions $\chi_\lambda^\pm$. From either of these we can calculate the function u, as

$$\chi_\lambda'\chi_\lambda^{-1} - \lambda\chi_\lambda J\chi_\lambda^{-1} = -\lambda J - JA. \tag{12.4}$$

In the next section we shall see that equation (12.3) always has a solution with the matrices $\chi_\lambda^\pm(x)$ invertible everywhere in their respective half-planes. For the moment, the important thing is that for such a solution the left-hand side of (12.4) is necessarily of the form $-\lambda J - JA$ for some off-diagonal matrix A independent of λ. For (12.3) implies that for real λ the left-hand side of (12.4) takes the same value whether we use $\chi_\lambda = \chi_\lambda^+$ or $\chi_\lambda = \chi_\lambda^-$, and it is therefore an *entire* matrix-valued function of λ with a simple pole with residue $-J$ at $\lambda = \infty$. This forces it to be $-\lambda J +$ (constant). I shall not discuss why the constant matrix must tend rapidly to zero as $x \to \pm\infty$. For that, see Faddeev and Takhtajan (1987).

Concerning the uniqueness of the solution, if $\widetilde{\chi}_\lambda^\pm$ is another solution of (12.3) we find

$$\widetilde{\chi}_\lambda^-(\chi_\lambda^-)^{-1} = \widetilde{\chi}_\lambda^+(\chi_\lambda^+)^{-1} \tag{12.5}$$

[7] I shall only consider the case when a_λ does not vanish for real λ, so that S_λ is defined and invertible. In contrast to the situation of Section 5, this need not be true.

for $\lambda \in \mathbb{R}$. Now the left-hand side of (12.5) is a meromorphic matrix-valued function in the lower half-plane, and tends to 1 as $\lambda \to \infty$. The right-hand side behaves similarly in the upper half-plane. Both sides are therefore equal to a function which is meromorphic, and hence *rational*, in the whole Riemann sphere. In other words, we have

Proposition 12.6 *The solution of (12.3), if it exists, is unique up to replacing $\chi_\lambda^\pm$ by $\rho_\lambda \chi_\lambda^\pm$, where ρ_λ is a rational matrix-valued function which is unitary on the real axis and such that $\rho_\infty = 1$. In particular, if $\chi_\lambda^\pm$ are invertible in their respective half-planes, then they are uniquely determined by (12.3).*

The unique solution of (12.3) with invertible $\chi_\lambda^\pm$ corresponds to a 'purely dispersive' choice of A. The others obtained from it by means of rational functions ρ_λ have 'solitons' on top of the dispersive background. We shall study them in Section 14.

13 Loop groups and the restricted Grassmannian

We have seen that inverting the scattering transformation depends on solving the following problem, sometimes called a *Riemann–Hilbert problem*:

> Given a smooth map $g : \mathbb{R} \to GL_n\mathbb{C}$ such that $g(\lambda) \to 1$ rapidly as $|\lambda| \to \infty$, to factorize it as $g = g_- g_+$, where g_+ and g_- are the boundary values of holomorphic maps $g_\pm : U_\pm \to GL_n\mathbb{C}$ defined in the upper and lower complex half-planes $U_\pm$, and $g_\pm \to 1$ as $\lambda \to \infty$ in $U_\pm$.

The factorization problem amounts to solving an integral equation. For if we write $g = 1 + \gamma$, $(g_+)^{-1} = 1 + \gamma_+$, and $g_- = 1 + \gamma_-$, then $g_- = g(g_+)^{-1}$ is equivalent to

$$\gamma_- = \gamma + \gamma_+ + \gamma\gamma_+ .$$

Taking the Fourier transform of this, and observing that $\hat{\gamma}_+$ and $\hat{\gamma}_-$, defined by

$$\hat{\gamma}_\pm(x) = \int \gamma_\pm(\lambda) e^{-i\lambda x} \, \mathrm{d}\lambda$$

vanish when $x \leqslant 0$ and $x \geqslant 0$ respectively, we have

$$\hat{\gamma}(x) + \hat{\gamma}_-(x) + \int_0^\infty \hat{\gamma}(x-y)\widehat{\gamma}_-(y)\,\mathrm{d}y = 0 \qquad (13.1)$$

for $x \geqslant 0$. This is an integral equation for $\hat{\gamma}_+$ when $\hat{\gamma}$ is given. In its application to inverse scattering, it is called the *Gelfand–Levitan–Marchenko* equation.[8]

It is more convenient to discuss loops $g : S^1 \to GL_n\mathbb{C}$ parametrized by the unit circle

$$S^1 = \{z \in \mathbb{C} : |z| = 1\}$$

rather than the real axis in the Riemann sphere S^2. This is, of course, only a notational change. We shall write $D_\pm$ for the two closed hemispheres of S^2 separated by S^1, with

$$D_+ = \{z \in \mathbb{C} : |z| \leqslant 1\}\,.$$

We shall now look for a factorization $g = g_-g_+$, where $g_\pm : D_\pm \to GL_n\mathbb{C}$ are smooth, and holomorphic in the interiors of the discs. I shall write $\mathcal{L}^\pm GL_n\mathbb{C}$, or just $\mathcal{L}^\pm$, for the corresponding subgroups of the group $\mathcal{L}GL_n\mathbb{C}$ of all smooth $g : S^1 \to GL_n\mathbb{C}$.

A loop g can be taken as the attaching map for a holomorphic vector bundle E on S^2 with fibre $\mathbb{C}^n$: we attach $D^- \times \mathbb{C}^n$ to $D^+ \times \mathbb{C}^n$ by

$$(z, \xi) \sim (z, g(z)\xi)$$

when $|z| = 1$. (If g is real analytic, so that it extends to a holomorphic map defined in a neighbourhood of S^1 in S^2, this construction obviously gives us a holomorphic vector bundle. It is true, but not obvious, that it does so for any smooth g.) A factorization $g_- = g(g_+)^{-1}$ is then the same thing as a holomorphic isomorphism $S^2 \times \mathbb{C}^n \to E$, i.e. a holomorphic trivialization of E. Not every bundle is trivial, and therefore the factorization is not always possible. In the accompanying lectures by Nigel Hitchin it was proved that any holomorphic bundle on S^2 breaks up as a sum

$$E \cong L^{k_1} \oplus \cdots \oplus L^{k_n}\,,$$

[8] This is an oversimplification. See Section II.4 of Faddeev and Takhtajan (1987).

where L^k is the kth tensor power of a basic line bundle L on S^2. The attaching function for L is simply z, so for $L^{k_1} \oplus \cdots \oplus L^{k_n}$ it is the diagonal loop

$$z^{\mathbf{k}} = \begin{pmatrix} z^{k_1} & & & \\ & z^{k_2} & & \\ & & \ddots & \\ & & & z^{k_n} \end{pmatrix}.$$

In other words, we have

Theorem 13.2. (Birkhoff's theorem) *Any loop g in GL_n can be factorized*

$$g = g_- z^{\mathbf{k}} g_+$$

for some multi-index $\mathbf{k}$, *where $g_\pm$ belongs to $\mathcal{L}^\pm GL_n\mathbb{C}$.*

The multi-index $\mathbf{k}$ which occurs here is uniquely determined up to reordering. A loop g has a *winding number* $\deg(g)$, which is defined as the winding number of $\det(g) : S^1 \to \mathbb{C}^\times$. We have

$$\deg(g) = \deg(z^{\mathbf{k}}) = \Sigma k_i,$$

for loops which extend over $D_\pm$ are null-homotopic. The connected component of the identity in $\mathcal{L}GL_n\mathbb{C}$ consists of the loops with winding number zero. It is stratified by subspaces $\mathcal{L}_{\mathbf{k}}$ corresponding to the multi-indices with $k_1 \geqslant k_2 \geqslant \cdots \geqslant k_n$ and $\Sigma k_i = 0$, and $\mathcal{L}_{\mathbf{k}}$ has complex codimension

$$\sum_{i<j} \max(k_i - k_j - 1,\ 0)\,.$$

Thus factorization $g = g_- g_+$ is possible if $\deg(g) = 0$ and g does not lie in a certain complex hypersurface in $\mathcal{L}GL_n\mathbb{C}$.

To prove Theorem 13.2 it is helpful to understand its relationship to a trivial factorization theorem for an element g of the group $GL_n\mathbb{C}$.

Proposition 13.3 *We can write*

$$g = \begin{pmatrix} a & b \\ c & d \end{pmatrix} = \begin{pmatrix} 1 & 0 \\ ca^{-1} & 1 \end{pmatrix} \begin{pmatrix} 1 & b \\ 0 & d - ca^{-1}b \end{pmatrix}$$

where the matrices are partitioned $(k + (N-k)) \times (k + (N-k))$, *if and only if* $\det(a) \neq 0$.

To view this result geometrically, consider the action of $GL_N\mathbb{C}$ on the Grassmanian manifold $Gr_k(\mathbb{C}^N)$ of all k-dimensional subspaces of $\mathbb{C}^N$. Let us write $\mathbb{C}^N = P \oplus Q$, where P and Q are spanned by the first k and last $N-k$ standard basis vectors of $\mathbb{C}^N$. The stabilizer of $P \in Gr_k(\mathbb{C}^N)$ is the subgroup G^+ of matrices of the form $\begin{pmatrix} * & * \\ 0 & * \end{pmatrix}$. But for any $g \in GL_N\mathbb{C}$, if gP is transversal to Q (i.e. $gP \cap Q = 0$) then gP is the graph of a map $w : P \to Q$. So

$$gP = \begin{pmatrix} 1 & 0 \\ w & 1 \end{pmatrix} P.$$

Thus $g^{-1}\begin{pmatrix} 1 & 0 \\ w & 1 \end{pmatrix}$ stabilizes P, and therefore belongs to G^+, giving us the factorization.

To adapt this argument to the loop group, we let $\mathcal{L}GL_n\mathbb{C}$ act in the obvious way on the Hilbert space $\mathcal{H}$ of L^2 functions $S^1 \to \mathbb{C}^n$. We then consider the *restricted Grassmannian* Gr_{res} consisting of all closed subspaces of $\mathcal{H}$ which are 'close', in a certain sense, to the subspace $\mathcal{H}_+$ consisting of the boundary values of all holomorphic maps $D_+ \to \mathbb{C}^n$. (If $\{e_i\}$ is the usual basis of $\mathbb{C}^n$, then the elements $e_i z^k$ for $k \geqslant 0$ form an orthonormal basis for $\mathcal{H}_+$.) A subspace W is said to be 'close' to $\mathcal{H}_+$ if the operator of orthogonal projection on to W differs from projection on to $\mathcal{H}_+$ by an integral operator with smooth kernel. The loop group $\mathcal{L}GL_n\mathbb{C}$ acts on Gr_{res}, and the stabilizer of $\mathcal{H}_+$ is plainly $\mathcal{L}^+$. The key to obtaining theorems about loop groups from the restricted Grassmannian is

Proposition 13.4 *A subspace $W \in Gr_{\text{res}}$ is of the form $g\mathcal{H}_+$ for some $g \in \mathcal{L}GL_n\mathbb{C}$ if and only if $zW \subset W$. In that case, W/zW has dimension n, and if $w_1, \ldots, w_n$ is an orthonormal basis of $W \ominus zW$ then the matrix w with columns $(w_1 \cdots w_n)$ belongs to $\mathcal{L}U_n$, and $W = w\mathcal{H}_+$.*

In proving Proposition 13.4 the main step is to show that if $w_1, \ldots, w_n$ is an orthonormal basis of $W \ominus zW$ then the vectors $w_1(\zeta), \ldots, w_n(\zeta)$ are an orthonormal basis of $\mathbb{C}^n$ at each point ζ of the circle. That follows from the formula

$$\langle w_i(\zeta), w_j(\zeta)\rangle \;=\; \sum_{n\in\mathbb{Z}} \langle z^n w_i,\; w_j\rangle\, \zeta^n$$

relating the inner product in $\mathbb{C}^n$ to that in $\mathcal{H}$. This argument gives us at the same time another factorization theorem.

Proposition 13.5 *Any loop $g \in \mathcal{L}GL_n\mathbb{C}$ can be written $g = wg_+$, with $w \in \mathcal{L}U_n$ and $g_+ \in \mathcal{L}^+$.*

Because the intersection $\mathcal{L}U_n \cap \mathcal{L}^+$ consists of constant loops (by the maximum modulus principle applied to g^*g and $(g^*g)^{-1}$) we have $\mathcal{L}GL_n\mathbb{C}/\mathcal{L}^+ \cong \mathcal{L}U_n/U_n \cong \Omega U_n$, the based loops in U_n, giving

Corollary $\Omega U_n \cong \{W \in Gr_{\text{res}} : zW \subset W\}$.

Returning to the Birkhoff Theorem 13.2, let $\mathcal{H}_-$ be the closed subspace of $\mathcal{H}$ spanned by $\{e_i z^k\}$ for $k < 0$, so that $\mathcal{H} = \mathcal{H}_+ \oplus \mathcal{H}_-$. To factorize $g \in \mathcal{L}GL_n\mathbb{C}$ in the form $g = g_-g_+$ it is enough to show that $g\mathcal{H}_+ = g_-\mathcal{H}_+$, where g_- is a loop whose columns belong to $z\mathcal{H}_-$. This is true if and only if $g\mathcal{H}_+$ is transversal to $\mathcal{H}_-$ (i.e. if $g\mathcal{H}_+ \oplus \mathcal{H}_- = \mathcal{H}$, though the sum is not orthogonal), for then $V = g\mathcal{H}_+ \cap z\mathcal{H}_-$ has dimension n, and a basis for V will serve as the columns of g_-.

A generic space $W \in Gr_{\text{res}}$ of virtual dimension zero will be transversal to $\mathcal{H}_-$, and so a generic loop g of winding number zero can be expressed $g = g_-g_+$. That is all I shall say here about the proof of Theorem 13.2, as the non-generic case is not relevant to inverse scattering. It is more important to mention that the map $g \mapsto g_-$ is a *meromorphic* function on the loop group, with a pole along the hypersurface of non-factorizable loops: the behaviour is precisely analogous to that of the map

$$\begin{pmatrix} a & b \\ c & d \end{pmatrix} \mapsto \begin{pmatrix} 1 & 0 \\ ca^{-1} & 1 \end{pmatrix}$$

of Proposition 13.3.

There are many variants of Proposition 13.4. One which we shall need in the next section concerns the group $GL_n(\mathbb{C}(z))$ of invertible matrices with entries in the field $\mathbb{C}(z)$ of rational functions of z. This group acts on the vector space $\mathcal{F}$ of all meromorphic $\mathbb{C}^n$-valued functions on $\mathbb{C}$ with at most finitely many poles. Any $\varphi \in \mathcal{F}$ can be written $\varphi = p^{-1}f$ for some polynomial $p \in \mathbb{C}[z]$ and f in the space $\mathcal{F}_+$ of holomorphic maps $\mathbb{C} \to \mathbb{C}^n$. Associated with $\mathcal{F}$ is the *rational Grassmannian* $Gr_{\mathrm{rat}}^{(n)}$ consisting of subspaces $W \in \mathcal{F}$ such that $zW \in W$ and

$$p\mathcal{F}_+ \subset W \subset q^{-1}\mathcal{F}_+ \tag{13.6}$$

for some polynomials p, q. This Grassmannian would not change if we replaced $\mathcal{F}$ by $\mathbb{C}(z)^n$, i.e. it is the space of all $\mathbb{C}[z]$-submodules of rank n in $\mathbb{C}(z)^n$. That makes the following result fairly obvious.

Proposition 13.7 *We have*

$$\begin{aligned} Gr_{\mathrm{rat}}^{(n)} &\cong GL_n(\mathbb{C}(z))/GL_n(\mathbb{C}[z]) \\ &\cong GL_n^{(\infty)}(\mathbb{C}(z)), \end{aligned}$$

where $GL_n^{(\infty)}(\mathbb{C}(z))$ *consists of the elements which take the value 1 when* $z = \infty$.

Note This is a purely algebraic statement, with no topologies involved: indeed the whole group $GL_n(\mathbb{C}(z))$ has no sensible topology. But for our purposes the subgroup $GL_n^{\infty}(\mathbb{C}(z))$ should be given the topology coming from the Taylor series expansion at $z = \infty$. This corresponds to the topology on $Gr_{\mathrm{rat}}^{(n)}$ for which the finite dimensional Grassmannian $Gr(p, q)$ of all spaces W satisfying (13.6) has its usual compact topology, and depends continuously on the monic polynomials p, q.

There is a hermitian form $\mathcal{F} \times \mathcal{F} \to \mathbb{C}$ defined by

$$(\varphi_1, \varphi_2) \mapsto \frac{1}{2\pi \mathrm{i}} \int \langle \varphi_1(\bar{z}),\ \varphi_2(z) \rangle \,\mathrm{d}z\,, \tag{13.8}$$

where $\langle\ ,\ \rangle$ is the usual hermitian form on $\mathbb{C}^n$, and the integral is taken around a large circle containing all the poles of φ_1 and φ_2. The subspace $\mathcal{F}_+$ is maximal isotropic for this form, and so is $W = \gamma\mathcal{F}_+$ if $\gamma \in GL_n^{\infty}(\mathbb{C}(z))$ is unitary on the real axis. In fact $W \mapsto W^{\perp}$ defines an antiholomorphic involution on $Gr_{\mathrm{rat}}^{(n)}$ for which $(\gamma\mathcal{F}_+)^{\perp} = \bar{\gamma}^{-1}\mathcal{F}_+$, where $\bar{\gamma}(z) = \overline{\gamma(\bar{z})}$. We have

Proposition 13.9 *The subgroup* Γ *of* $GL_n^{(\infty)}(\mathbb{C}(z))$ *consisting of elements which are unitary on the real axis can be identified with*

$$\{W \in Gr_{\mathrm{rat}}^{(n)} : W^{\perp} = W\}\,.$$

A space $W \in Gr_{\mathrm{rat}}^{(n)}$ has a *support*, which is the smallest finite subset S of $\mathbb{C}$ such that (13.6) is true for some polynomials p, q whose roots lie in S. I shall write Gr_S for the set of W with support contained in S. It corresponds to the subgroup of $\gamma \in Gr_n^{(\infty)}(\mathbb{C}(z))$ such that both γ and γ^{-1} have their poles in S. By the Chinese remainder theorem we have

$$Gr_{\mathrm{rat}}^{(n)} = \prod_{\zeta \in \mathbb{C}}^{\mathrm{res}} Gr_\zeta$$

as sets, where $\prod^{\mathrm{res}}$ means that the product is restricted to families $\{W_\zeta\}$ for which $W_\zeta = \mathcal{F}_+$ for all but finitely many ζ. Each space Gr_ζ can be identified with the Laurent polynomial loop group of U_n. The correspondence in Proposition 13.7 does not respect the topology, for the topology of $Gr_{\mathrm{rat}}^{(n)}$ allows the support of a point W to move continuously with W. The correct picture of $Gr_{\mathrm{rat}}^{(n)}$ is as a 'labelled configuration space': a point consists of a finite subset $\zeta_1, \ldots, \zeta_k$ of distinct points of $\mathbb{C}$, each point ζ_i being 'labelled' with a point of Gr_{ζ_i}. The points ζ_i can 'collide', and then the labels are appropriately amalgamated.

If $W \in Gr_{\mathrm{rat}}^{(n)}$ has support S then $W^\perp$ has support $\bar{S}$. It follows that (as sets again)

$$\Gamma \quad \cong \quad \prod_{\mathrm{Im}(\zeta)>0}^{\mathrm{res}} Gr_\zeta \, .$$

In the next section we shall be interested in the sub-semigroup Γ^+ of Γ consisting of elements which have no poles in the upper half-plane. The elements of Γ^+ with support $\{\zeta, \bar{\zeta}\}$, where $\mathrm{Im}(\zeta) > 0$, form a semigroup Γ_ζ^+ which is described very explicitly in Segal (1981). It is the union of a sequence of connected components $\Gamma_{\zeta,k}^+$, each of which is a compact complex algebraic variety of complex dimension $k(n-1)$. The component $\Gamma_{\zeta,1}^+$ can be identified with $\mathbb{P}_{\mathbb{C}}^{n-1}$: it consists of elements γ_L of the form

$$\gamma_L(z) \quad = \quad (1 - P_L) + \left(\frac{z-\zeta}{z-\bar{\zeta}}\right) P_L, \qquad (13.10)$$

where P_L is orthogonal projection on to a line L in $\mathbb{C}^n$. The component $\Gamma_{\zeta,k}^+$ consists of all k-fold products

$$\gamma_{L_1} \gamma_{L_2} \cdots \gamma_{L_k} \, ,$$

and the semigroup is free except that

$$\gamma_L \, \gamma_{L'} \;=\; \gamma_{L \oplus L'}$$

depends only on $L \oplus L'$ if L and L' are orthogonal. (The element γ_L is traditionally called a *Blaschke factor.*)

Abstract scattering theory

We can define a restricted Grassmannian whenever we have a Hilbert space $\mathcal{H}$ and a closed subspace $\mathcal{H}_+$ of infinite dimension and codimension.

Suppose that $u : \mathcal{H} \to \mathcal{H}$ is a unitary operator. If $\mathcal{H}_+$ is any closed subspace such that $u(\mathcal{H}_+) \subset \mathcal{H}_+$ then the Hilbert space $l^2(\mathbb{Z}; V)$ of sequences $\{\xi_n\}_{n\in\mathbb{Z}}$ in $V = \mathcal{H} \ominus u(\mathcal{H})$ can be embedded in $\mathcal{H}$ by the isometry

$$\{\xi_n\} \mapsto \sum u^n(\xi_n)\,.$$

If $\mathcal{H}_+$ satisfies the additional conditions

$$\bigcap_{n\geqslant 0} u^n(\mathcal{H}_+) = 0$$

and

$$\bigcup_{n\leqslant 0} u^n(\mathcal{H}_+) \quad \text{is dense in } \mathcal{H}$$

then

$$l^2(\mathbb{Z}; V) \to \mathcal{H}$$

is an isomorphism, and makes the action of u on $\mathcal{H}$ correspond to the shift map on $l^2(\mathbb{Z}; V)$, and makes $\mathcal{H}_+$ correspond to the sequences $\{\xi_n\}$ such that $\xi_n = 0$ when $n < 0$. In this case $\mathcal{H}_+$ is called an *outgoing subspace.*

Because $l^2(\mathbb{Z}; V) \cong L^2(\mathbb{T}; V)$, with the shift map on l^2 becoming the operation M_z of multiplication by $z = \mathrm{e}^{\mathrm{i}\theta}$ on $\mathbb{T}$, the existence of an outgoing subspace means that the spectrum of u is precisely $\mathbb{T} \subset \mathbb{C}$ with uniform multiplicity $\dim(V)$. In other words, the existence of $\mathcal{H}_+$ together with the dimension of $\mathcal{H}_+/u(\mathcal{H}_+)$ completely determines u up to conjugation.

Given $\mathcal{H}_+$, we can define the restricted Grassmannian Gr_{res}, and we have

Proposition 13.11 $\{W \in Gr_{\text{res}} : u(W) \subset W\} \quad \cong \quad \Omega U(V)$.

If, in addition to an outgoing subspace $\mathcal{H}_+$ we are given an *incoming* subspace $\mathcal{H}_-$, i.e. one which would be outgoing when u is replaced by u^{-1}, then $\mathcal{H}_-^{\perp}$ is a second outgoing subspace, and we have two isomorphisms

$$S_{\pm} : L^2(\mathbb{T}; V) \to \mathcal{H},$$

each making u correspond to M_z. We call $S = (S_-)^{-1}S_+$ the *scattering operator*. The commutant of M_z is the group of measurable maps $\mathbb{T} \to U(V)$, so

$$S \in \mathcal{L}_{\text{meas}}U(V).$$

Evidently we have $S(\mathcal{H}_+) = \mathcal{H}_-^{\perp}$. The significance of the restricted Grassmannian Gr_{res}, defined in terms of $\mathcal{H}_+$, is that $\mathcal{H}_-^{\perp} \in Gr_{\text{res}}$ if and only if the loop S is smooth.

If we have a one-parameter unitary group

$$\{U_t : \mathcal{H} \to \mathcal{H}\}_{t \in \mathbb{R}}$$

instead of a single operator $u : \mathcal{H} \to \mathcal{H}$ we can define incoming and outgoing subspaces as before. We can write $U_t = \mathrm{e}^{\mathrm{i}tA}$ for some unbounded self-adjoint operator A. A subspace $\mathcal{H}_+$ is outgoing for $\{U_t\}$ if and only if it is outgoing for the unitary operator

$$u = (1 + \mathrm{i}A)(1 - \mathrm{i}A)^{-1},$$

for

$$u = \int_0^{\infty} (2U_t - 1)\mathrm{e}^{-t}\,\mathrm{d}t.$$

The existence of $\mathcal{H}_+$ now means that the spectrum of A is $\mathbb{R}$ with uniform multiplicity $V = \mathcal{H}_+ \ominus u(\mathcal{H}_+)$, and

$$\varphi \mapsto \int_{\mathbb{R}} U_t \hat{\varphi}(t)\,\mathrm{d}t$$

defines a canonical isomorphism

$$L^2(\mathbb{R}; V) \quad \to \quad \mathcal{H}$$

relating U_t to multiplication by $\mathrm{e}^{\mathrm{i}tx}$.

This is essentially the situation we described in Section 5, where the space of solutions of the wave equation—or, more precisely, the subspace orthogonal to the finite dimensional subspace of bound states—was decomposed by means of an incoming and an outgoing subspace. The abstract formulation just presented is due to Lax and Phillips (1967).

14 Integrable systems and the restricted Grassmannian

Up to this point we have, except in Section 4, always studied the KdV equation with specific boundary conditions in 'space'—i.e. we have considered either rapidly decreasing functions on the line, or else functions on the circle. We shall now adopt a quite different approach which is *local* in the space variable.

Suppose then that we have an operator $L_u = -(\mathrm{d}/\mathrm{d}x)^2 + u$ where u is defined just in some open interval I of $\mathbb{R}$. Let us look for a solution of $L_u \varphi_\lambda = \lambda^2 \varphi_\lambda$ which is of the form

$$\varphi_\lambda(x) = \mathrm{e}^{\mathrm{i}\lambda x}\left\{1 + \frac{a_1(x)}{\lambda} + \frac{a_2(x)}{\lambda^2} + \cdots\right\}. \tag{14.1}$$

We find that the functions a_k can be calculated iteratively from

$$\begin{aligned} 2ia_1' &= u\,, \\ 2ia_{k+1}' &= L_u a_k \quad \text{for } k \geqslant 1. \end{aligned} \tag{14.2}$$

Thus we can always find a formal power series (14.1) which is an eigenfunction of L_u. Each equation (14.2) involves a new constant of integration, and so φ_λ is unique only up to multiplication by a formal power series in λ^{-1} which is independent of x. For most choices of u the series (14.1) will diverge for all λ, as we shall see. Suppose, however, that for some u we have a formal Baker function φ_λ which converges for $|\lambda| \geqslant \mathbb{R}$. We shall assign to φ_λ a point W of the restricted Grassmanian of the Hilbert space $\mathcal{H}$ of L^2 functions $S_R \to \mathbb{C}$, where

$$S_R = \{\lambda \in \mathbb{C} : |\lambda| = R\}\,.$$

To do this, first fix a point $x \in I$, and consider the closed subspace W_x of $\mathcal{H}$ spanned by the sequence of functions of $\lambda \in S_R$

$$\varphi_\lambda(x),\ \varphi_\lambda'(x),\ \varphi_\lambda''(x),\ \varphi_\lambda'''(x),\ \ldots\ , \tag{14.3}$$

or, equally well, by $\lambda^{2k}\varphi_\lambda(x)$ and $\lambda^{2k}\varphi'_\lambda(x)$ for $k \geqslant 0$. I shall omit the proof that W_x does belong to Gr_{res}, which follows from the second description (see Segal and Wilson 1985, p. 34). From the first description of W_x the x-derivative of any generator of W_x is contained in W_x, giving us

Proposition 14.4 *The space W_x is independent of x.*

Notice, however, that $W = W_x$ depends on the choice of the convergent Baker function φ_λ, and not just on L_u. Changing the choice of φ_λ changes W to γW, where $\gamma = \sum_{k \geqslant 0} c_k \lambda^{-k}$ is a holomorphic map $D_R^- \to \mathbb{C}^\times$.

The space W clearly satisfies $\lambda^2 W \subset W$. Conversely,

Proposition 14.5 *If $W \in Gr_{\text{res}}$ satisfies $\lambda^2 W \subset W$ then for all $x \in \mathbb{C}$ such that $\mathrm{e}^{-\mathrm{i}\lambda x}W$ is transversal to $\mathcal{H}_-$ there is a unique element $\varphi_\lambda(x)$ of W of the form 14.1. Furthermore $\varphi_\lambda(x)$ is meromorphic in x for all $x \in \mathbb{C}$, and satisfies $L_u\varphi_\lambda = \lambda^2\varphi_\lambda$, where $u = 2\mathrm{i}a'_1$.*

Here, of course, we are writing $\mathcal{H} = \mathcal{H}_+ \oplus \mathcal{H}_-$, were $\mathcal{H}_+$ and $\mathcal{H}_-$ are spanned by λ^k for $k \geqslant 0$ and $k < 0$ respectively. The proof of the existence of $\varphi_\lambda(x)$ is obvious: if $\mathrm{e}^{-\mathrm{i}\lambda x}W$ is transversal to $\mathcal{H}_-$ then $\mathrm{e}^{-\mathrm{i}\lambda x}W$ and $1 + \mathcal{H}_-$ meet in a single point. To see that $L_u\varphi_\lambda = \lambda^2\varphi_\lambda$ we simply observe that for each x the function

$$\mathrm{e}^{-\mathrm{i}\lambda x}(L_u\varphi_\lambda - \lambda^2\varphi_\lambda)$$

of λ belongs to $\mathrm{e}^{-\mathrm{i}\lambda x}W$ but also to $\mathcal{H}_-$. Finally, the meromorphicity follows from the remark after the Corollary to Proposition 13.5 (see Segal and Wilson 1985, page 51).

We now see that a convergent Baker function cannot exist unless u extends to a meromorphic function on all of $\mathbb{C}$. If, for example, u is rapidly decreasing on the line then we know from Section 5 that there are unique eigenfunctions $\widetilde{\varphi}_\lambda^+$ and φ_λ^- of the form (14.1) which are holomorphic for λ in the upper and lower half-planes respectively, while on the real axis $\varphi_\lambda^+ - \varphi_\lambda^-$ is the reflection coefficient R_λ. In this case, therefore, a convergent Baker function can exist for $|\lambda| \geqslant R$ only if the reflection coefficient has compact support. The reflection coefficient is a 'non-linear Fourier transform' of u, and the Fourier transform of a function with compact support is entire, so it is perhaps not surprising that u is meromorphic when R_λ has compact support.

The next point to understand is that points W of Gr_{res} such that $\lambda^2 W \subset W$ encode not only operators L_u, but in fact complete solutions to the KdV equation and the whole system of commuting flows associated with it. For, just as we associated $\varphi_\lambda(x)$ to W in Proposition 14.5, we can show that for a generic sequence $\mathbf{x} = (x_1, x_2, x_3, \ldots)$ there is a unique element $\varphi_\lambda(\mathbf{x})$ of W of the form

$$\varphi_\lambda(\mathbf{x}) = \mathrm{e}^{\mathrm{i}\Sigma x_k \lambda^k} \left\{ 1 + \frac{a_1(\mathbf{x})}{\lambda} + \frac{a_2(\mathbf{x})}{\lambda^2} + \cdots \right\}.$$

(Of course, we need $f = \Sigma x_k \lambda^k$ to converge for $|\lambda| = R$.) If we observe that

$$\frac{\partial \varphi_\lambda}{\partial x_k} = \mathrm{e}^{\mathrm{i}f} \{ \mathrm{i}\lambda^k + O(\lambda^{-1}) \},$$

while

$$\frac{\partial^m \varphi_\lambda}{\partial x_1^m} = \mathrm{e}^{\mathrm{i}f} \{ (\mathrm{i}\lambda)^m + O(\lambda^{m-2}) \},$$

we can easily show that there are unique ordinary differential operators P_k^+ of the form

$$P_k^+ = \left(\frac{\partial}{\partial x}\right)^k - \mathrm{i}k a_1' \left(\frac{\partial}{\partial x}\right)^{k-2} + \cdots,$$

where $x = x_1$, such that

$$\frac{\partial \varphi_\lambda(\mathbf{x})}{\partial x_k} = P_k^+ \phi_\lambda(\mathbf{x}).$$

In particular, if we take $x_3 = t$ we find that P_3^+ is the P occurring in the Lax form of the KdV equation, so that $u = 2\mathrm{i}a_1'(x, t, 0, \ldots)$ is a solution of the KdV equation. But, more generally, we easily see that P_k^+ is the 'fractional power' $(L_u^{k/2})_+$ discussed in Section 4.

If we write

$$Gr_{\mathrm{res}}^{(2)} = \{ W \in Gr_{\mathrm{res}} : \lambda^2 W \subset W \},$$

what we have now proved is that there is a map

$$Gr_{\text{res}}^{(2)}/\mathcal{L}^{-}\mathbb{C}^{\times} \longrightarrow \{\text{meromorphic functions } u\}$$

such that the action of the abelian group $\mathcal{L}^{+}\mathbb{C}^{\times}$ on the left corresponds to the hierarchy of commuting KdV flows on the right—to be precise, $\gamma = e^{\Sigma x_k \lambda^k}$ in $\mathcal{L}^{+}\mathbb{C}^{\times}$ corresponds to flowing for 'time' x_k along the kth flow. (Because $\lambda^2 W \subset W$, the space γW is independent of the x_k with k even, so only the odd flows are non-trivial.)

The space $Gr_{\text{res}}^{(2)}$ can itself be identified with the based loop-group ΩU_2 of U_2. This follows from the discussion at the end of Section 13, for $\mathcal{H} = L^2(S_R; \mathbb{C})$ is a Hilbert space equipped with a unitary transformation $f \mapsto \lambda^2 f$ such that $\lambda^2 \mathcal{H}_+ \subset \mathcal{H}_+$ with codimension 2. We can also say

$$Gr_{\text{res}}^{(2)} \cong \mathcal{L}GL_2\mathbb{C}/\mathcal{L}^{+}GL_2\mathbb{C}.$$

The action of $\mathcal{L}\mathbb{C}^{\times}$ on $Gr_{\text{res}}^{(2)}$ comes from the inclusion of $\mathcal{L}\mathbb{C}^{\times}$ in $\mathcal{L}GL_2\mathbb{C}$ by

$$\gamma \mapsto \begin{pmatrix} a & \lambda^{-1/2}b \\ \lambda^{1/2}b & a \end{pmatrix},$$

where $a = \frac{1}{2}(\gamma(\lambda^{1/2}) + \gamma(\lambda^{-1/2}))$ and $b = \frac{1}{2}(\gamma(\lambda^{1/2}) - \gamma(\lambda^{-1/2}))$. If we examine this construction we find that the space of meromorphic solutions of the KdV hierarchy we have found is parametrized precisely by the based loop space ΩS^2.

The NLS equation

The local theory of the KdV equation has a precise parallel for the NLS equation. I shall carry it a little further to show how it can be applied to study the rapidly decreasing solutions on the line.

The equation

$$\left(J\frac{\mathrm{d}}{\mathrm{d}x} + A\right)\varphi = \lambda\varphi$$

has formal solution matrices $M_\lambda(x) = \chi_\lambda(x)\mathrm{e}^{-\lambda J x}$, where

$$\chi_\lambda(x) \sim 1 + \lambda^{-1}a_1(x) + \lambda^{-2}a_2(x) + \cdots . \tag{14.6}$$

This solution is unique up to multiplication on the right by a *diagonal* matrix-valued function of λ, independent of x.

If the series (14.6) converges for $|\lambda| \geqslant R$ then for each x the *rows* of the 2×2 matrix $M_\lambda(x)$ are elements of the Hilbert space $\mathcal{H}$ of L^2 functions on $|\lambda| = R$ with values in $\mathbb{C}^2$. These rows and their successive x-derivatives span a subspace W_x belonging to the restricted Grassmannian of $\mathcal{H}$, just as in 14.3. Once again $W = W_x$ is independent of x, and it satisfies $\lambda W \subset W$. Everything proceeds as for the KdV equation.

Now suppose that A is rapidly decreasing. We have the solutions $M_\lambda^\pm$ for λ in the two half-planes, related by

$$M_\lambda^- = M_\lambda^+ S_\lambda$$

when λ is real, as well as by the unitarity relation (12.2). If there is no dispersive scattering then $S_\lambda = 1$, and $M_\lambda^\pm$ fit together to define a meromorphic function in all of $\mathbb{C}$, and the space W corresponds to the rational loop $\rho_\lambda = \chi_\lambda(0)$ in $GL_2(\mathbb{C})$, and belongs to the Grassmannian denoted $Gr_{\mathrm{rat}}^{(2)}$ in Section 13. We also know that $\rho_\infty = 1$, and that ρ_λ has no poles in the upper half-planes, and is unitary on the real axis. Such rational loops form a semigroup Γ^+, whose connected components Γ_n^+ correspond to the number n of zeros of $a_\lambda = \det(M_\lambda)$ in the upper half-plane. They correspond to a subspace Gr^+ of the rational Grassmannian. The component Γ_n^+ is a compact complex algebraic variety of complex dimension $2n$. As x traverses the line $\mathbb{R}$ the rational loop $\chi_\lambda(x)$ evolves in correspondence with the subspace $W\mathrm{e}^{\lambda J x} \in Gr^+$, in other words according to the flow of Gr^+ induced by the flow $\{\mathrm{e}^{\lambda J x}\}$ on $\mathcal{F}$. We can decompose the isotropic Grassmannian as

$$\prod_{\mathrm{Im}(\zeta)>0} Gr_\zeta^+$$

as in §13. The flow $\{\mathrm{e}^{\lambda J x}\}$ leaves the support of each element of Gr^+ fixed, and acts independently on each Gr_ζ^+. The same applies to the flow generated by $\{\mathrm{e}^{\lambda^2 J t}\}$, which corresponds to letting the solution evolve for time t according to the NLS equation. So the space of solutions breaks up into a product of pieces labelled by numbers ζ in the upper half-plane which determine the velocity and frequency of oscillation of the soliton.

The fixed points of $e^{\lambda J x}$ on Gr_ζ^+ are the discrete set $W_{k,m}$, for $k, m \geqslant 0$, where $W_{k,m}$ corresponds to the loop

$$\begin{pmatrix} f^k & 0 \\ 0 & f^m \end{pmatrix}, \quad \text{with} \quad F = \frac{\lambda - \zeta}{\lambda - \bar{\zeta}}.$$

These all correspond to the trivial potential $A = 0$. (Recall that the solutions $M_\lambda^\pm(x)$ were determined by A only up to right multiplication by a diagonal function of λ.) For a general point W of Gr_ζ^+ the trajectory $We^{\lambda J x}$ goes from $W_{k,m}$, to $W_{k+r,\, m-r}$ as $-\infty < x < \infty$, for some integer $r \geqslant 0$. These trajectories sweep out a space X_ζ^r which is clearly independent of (k, m), and precisely parametrize the r-fold multisolitons of type ζ. This gives us a rather complete picture of the space of dispersionless potentials A.

If there is dispersive scattering, then the loop S_λ is, as we saw in Section 12, the attaching function for a holomorphic bundle E on the Riemann sphere $S^2 = \mathbb{C} \cup \{\infty\}$, and the rows of $M_\lambda^\pm(x)$ are meromorphic sections of $E|\mathbb{C}$. Because S_λ is hermitian the bundle E has a hermitian structure in the sense that the fibre $E_{\bar{\lambda}}$ is dual to $\bar{E}_\lambda$. Let us write $\mathcal{F}_E$ for the space of meromorphic sections of $E|\mathbb{C}$, and Gr_E for the associated rational Grassmannian of subspaces $W \subset \mathcal{F}_E$ such that $\lambda W \subset W$ and

$$p\mathcal{F}_E^+ \quad \subset \quad W \quad \subset \quad q^{-1}\mathcal{F}_E^+$$

for some polynomials $p, q \in \mathbb{C}[\lambda]$, where $\mathcal{F}_E^+$ denotes the holomorphic sections of $E|\mathbb{C}$. Inside Gr_E there is a subspace Gr_E^+ consisting of those W such that $W^\perp = W$ and $W \subset \mathcal{F}_E^{\mathrm{UHP}}$, where $\mathcal{F}_E^{\mathrm{UHP}}$ is the sections which are holomorphic in the upper half-plane.

We now have a precise analogue of the theory described above for dispersionless potentials. The subspace W_x of Gr_E spanned by the rows of $M_\lambda^\pm(x)$ and their x-derivatives is independent of x, and belongs to Gr_E^+. Conversely, for any $W \in Gr_E^+$ we can find a potential A for which $M_\lambda^\pm$ generate W. In fact we shall have

$$M_\lambda^\pm(X) = \rho_\lambda(x)\widetilde{M}_\lambda^\pm(x),$$

where $\widetilde{M}_\lambda^\pm$ is the purely dispersive solution with scattering loop S_λ described in Section 12, and $\rho_\lambda(x)$ belongs to the subsemigroup Γ^+ of $Gl_2(\mathbb{C}(\lambda))$, and becomes diagonal as $x \to \pm\infty$. All of this is described very fully and explicitly in Chapter II of Faddeev and Takhtajan (1987).

15 Algebraic curves and the Grassmannian

The construction just described of meromorphic solutions to the KdV equation from the Grassmannian is not directly practical. A source of explicit spaces $W \in Gr_{\text{res}}$ is provided by algebraic curves.

Let Σ be an algebraic curve or compact Riemann surface equipped with

(i) a distinguished point P and a local parameter λ at P which identifies a closed neighbourhood D_Σ of P with the disc

$$D_- = \{\lambda \in \mathbb{C} : |\lambda| \geqslant R\}$$

in the Riemann sphere, so that $\lambda(P) = \infty$,

(ii) a holomorphic line bundle L on Σ with a given trivialization of $L|D_\Sigma$.

Then we define W as the space of boundary values of holomorphic sections of L defined outside D_Σ, i.e. $W = \Gamma(L|\Sigma_0)$, where $\Sigma_0 = \Sigma -$ (interior of D_Σ). Here sections of L over the boundary of Σ_0 are identified with complex-valued functions on the circle S_R by the given trivialization of $L|D_\Sigma$. I shall omit the proof that W does belong to Gr_{res} (see Pressley and Segal 1986, p. 159). Notice that changing the trivialization of $L|D_\Sigma$ changes W by multiplication by an element of $\mathcal{L}^-\mathbb{C}^\times$, exactly the same ambiguity that we encountered in §14 in associating a space $W \in Gr_{\text{res}}$ to an operator L_u.

The choice of the curve Σ and the local parameter λ gives us a surjective homomorphism

$$j_P : \mathcal{L}\mathbb{C}^\times \to \text{Jac}\,(\Sigma)$$

from the loop group to the Jacobian variety of the line bundles on Σ (which is a group under $\otimes$): for to a loop γ we can associate a line bundle L_γ got by attaching the trivial bundles $\Sigma_0 \times \mathbb{C}$ and $D_\Sigma \times \mathbb{C}$ by means of γ. Any holomorphic line bundle L on Σ can be constructed in this way, for $L|\Sigma_0$ and $L|D_\Sigma$ are necessarily trivial.

Proposition 15.1 *If W arises as described from the line bundle L then γW arises from $L \otimes L_\gamma$, with its trivialization over D_Σ induced from that of L.*

For W to correspond to an operator L_u as in Section 14 we need, among other things, that $\lambda^2 W \subset W$. This is true if (and in fact only if) $\lambda^2 : D_\Sigma \to S^2$ extends to a holomorphic $f : \Sigma \to S^2$ such that $f^{-1}(D_+) = \Sigma_0$, i.e. if Σ is a ramified double cover of S^2, with $\infty \in S^2$ as one of the branch points. In particular, Σ must be hyperelliptic.

We need two additional conditions on W. First, if W is transversal to $\mathcal{H}_-$ it must have virtual dimension 0. Using the Riemann–Roch theorem for Σ we readily find

Proposition 15.2 *The virtual dimension of* W *is* $\deg(L)-g$, *where* g *is the genus of* Σ.

Thus we need $\deg(L) = g$. Then W is transversal to $\mathcal{H}_-$ if $W \cap \mathcal{H}_- = 0$. Now $W \cap \mathcal{H}_-$ is the space of sections over Σ of $L \otimes L_{z^{-1}}$. This bundle is conventionally denoted $L(-1)$—for $L_{z^{-1}}$ is the bundle whose sections are the sheaf of holomorphic functions which vanish at $P \in \Sigma$—so the condition we need is

$$H^0(\Sigma; L(-1)) = 0\,,$$

just as in Hitchin's lectures.

In Section 14 we saw that the hierarchy of KdV flows corresponded to moving $W \in Gr_{\text{res}}$ by $W \mapsto e^{\Sigma x_k \lambda^k} W$. So Proposition 15.1 shows that each orbit of the KdV flows on the space of meromorphic functions u produced by a curve Σ is a copy of the degree g component of the Jacobian of Σ, on which the flows are the action of the identity component $\text{Jac}_0(\Sigma)$. In fact every finite dimensional orbit coming from the Grassmannian arises from an algebraic curve in this way.

One further thing should be said to complete the dictionary between the loop group picture and Hitchin's. In his picture the solutions are constructed from the direct image vector bundle f_*L on S^2. Such a bundle can be constructed by means of an attaching function in $\mathcal{L}GL_2\mathbb{C}$, and the space

$$Gr^{(2)}_{res} \cong \mathcal{L}GL_2\mathbb{C}/\mathcal{L}^+GL_2\mathbb{C}$$

can be identified with the set of holomorphic bundles of rank two on S^2 which are trivialized over D_-. In this correspondence the space W corresponds to the bundle f_*L, and the inclusion $\mathcal{L}\mathbb{C}^\times \to \mathcal{L}GL_2\mathbb{C}$ described in Section 14 takes the attaching map of L to that of f_*L.

Bibliography

Akhiezer, N. I. (1965). *The classical moment problem.* Oliver and Boyd, Edinburgh.

Faddeev, L. D. (1962). The inverse problem of the quantum theory of scattering. *J. Math. Phys.*, **4**, 72–104.

Faddeev, L. D. and Takhtajan, L. A. (1987). *Hamiltonian methods in the theory of solitons.* Springer, Berlin.

Gelfand, I. M. and Dikii, L. A. (1976). Fractional powers of operators and Hamiltonian systems. *Funct. Anal. Appl.*, **10**(4), 13–29 (Russian), 259–73 (English).

Lax, P. D. and Phillips, R. S. (1967). *Scattering theory.* Academic Press, New York.

Moser, J. (1975). Finitely many mass points on the line under the influence of an exponential potential. Springer Lecture Notes in Physics, vol. 38, 467–497. Springer-Verlag, Berlin.

Moser, J. (1979). Geometry of quadrics and spectral theory. In *The Chern Symposium* (ed. W. Y. Hsiang *et al*), 147–88. Springer-Verlag, Berlin.

McKean, H. P. and Trubowitz, E. (1976). Hill's operator and hyperelliptic function theory in the presence of infinitely many branch points. *Comm. Pure Appl. Math.*, **29**, 143–226.

Newell, A. C. (1983). The history of the soliton. *J. Appl. Mech.*, **105**, 1127–37.

Pressley, A. N. and Segal, G. B. (1986). *Loop groups.* Oxford University Press.

Segal, G. B. (1981). Unitary representations of some infinite dimensional groups. *Commun. Math. Phys.*, **80**, 301–42.

Segal, G. B. (1989). Loop groups and harmonic maps. In *Advances in Homotopy Theory* (ed. S. M. Salamon, B. F. Steer, and W. A. Sutherland). Lond. Math. Soc. Lecture Note Series, vol. 139.

Segal, G. B. (1991). The geometry of the KdV equation. *Int. J. Mod. Phys.*, **A6**, 2859–69.

Segal, G. B. and Wilson, G. (1985). Loop groups and equations of KdV type. *Publ. Math. I.H.E.S. Paris*, **61**, 5–65.

Uhlenbeck, K. (1989). Harmonic maps into Lie Groups (classical solutions of the chiral model). *J. Diff. Geom.*, **30**, 1–50.

Wilson, G. (1979). Commuting flows and conservation laws for

Lax equations. *Math. Proc. Camb. Phil. Soc.*, **86**, 131–43.

4

Integrable Systems and Twistors

Richard Ward

Notes by Martin Speight

1 General comments on integrable systems

For the purpose of this chapter, integrable systems are special kinds of non-linear differential equations (although one should note that there are also integrable difference equations, and the study of these is increasingly prominent). For essentially any system of differential equations, one has local existence theorems, so existence of solutions is not the issue. The question, rather, is what the solutions are like. For a non-linear differential equation chosen at random, one has no hope of writing down explicit solutions in terms of known functions. The classification of a function as 'known' may seem artificial, having more to do with what we've bothered to learn than the function itself; but in fact there is more to it than that. The known functions (e.g. trigonometric functions or elliptic functions) are known because they have nice properties, and can be defined using only a finite amount of information. The functions involved in the solutions of a generic non-linear differential equation are so awful as to be literally indescribable in any but a tautological way (that is, one can really only define them as solutions of the differential equation). An extreme example of this is chaotic systems. At the opposite end of the scale, some equations can effectively be solved explicitly: there are dense families of explicit solutions involving known functions. This gives a rough definition of integrable systems (there are several definitions which are less vague, but they tend to be too restrictive).

Integrable systems are extremely special (almost every non-linear differential equation is not integrable). Nevertheless, they are important:

- They are mathematically beautiful, with links to geometry, etc.
- Many of the nice properties of solutions of integrable systems are more robust than integrability itself. For example, solitary waves in a shallow water channel are *approximated* by an integrable system, the Korteweg–de Vries equation. The full fluid-dynamical evolution equation of the system is not integrable, yet it shares some of the nice properties (solitons, stability) of its integrable approximation. The same is true of light pulses in optical fibres, which are approximately solutions of the non-linear Schrödinger equation.
- Even systems which are far from integrable may have an integrable 'heart' which tells one much about their behaviour. For example, one can obtain a good picture of the solutions of the time-dependent BPS monopole system (which is not integrable), by understanding the static system (which is).

2 Some elementary geometry

Consider a m-dimensional manifold M with local coordinates

$$x^1, x^2, \ldots, x^m$$

(for our purposes it suffices to take $M = \mathbb{R}^m$). A *vector bundle* of rank p over M is a structure in which a p-dimensional vector space V_x is attached to each point $x \in M$. A *vector field* is a smooth allocation of a vector $v(x) \in V_x$ to each point x. In concrete terms, we can choose a basis for each V_x, so that $v(x)$ is a column p-vector of functions of x^a. The choice of basis is often called a choice of *gauge*. A guiding principle is that all constructions should be independent of the choice of basis, that is, *gauge-invariant*. A *gauge transformation* (smooth pointwise change of basis) acts on the column vector v by

$$v(x) \mapsto \widehat{v}(x) = \Lambda(x)v(x), \tag{2.1}$$

where Λ is a non-singular $p \times p$ matrix of functions of x.

How does one make sense of the idea that $v(x)$ is constant in the x^1 direction (say)? The naive answer

$$\frac{\partial v}{\partial x^1} = 0 \tag{2.2}$$

is not gauge-invariant. To get something which is, one has to introduce a *connection* on the vector bundle. Let us use the shorthand notation $\partial_a = \partial/\partial x^a$. The constancy condition is $D_1 v = 0$, where

$$D_a v = \partial_a v + A_a v \tag{2.3}$$

is the *covariant derivative* of v. Here A_a is a $p \times p$ matrix of functions of x which transforms under gauge transformations as

$$A_a \mapsto \widehat{A}_a = \Lambda A_a \Lambda^{-1} - (\partial_a \Lambda)\Lambda^{-1}. \tag{2.4}$$

Then the covariant derivative transforms as $D_a v \mapsto \Lambda D_a v$ (so $D_a v$ transforms in the same way as v itself); the crucial point is that its *vanishing* is gauge-invariant. If $D_1 v = 0$, then one says that v is covariantly constant in the x_1-direction. In physics, A_a is called a *gauge potential*. Generally, Λ takes values in a Lie group G, and A_a takes values in the corresponding Lie algebra $\mathfrak{g}$. Assuming that V_x are real vector spaces, the gauge group we've been using so far is $GL(p, \mathbb{R})$. Interesting examples often involve complex vector bundles, and compact subgroups of the general linear group such as $SU(p)$.

If we demand that $v(x)$ be covariantly constant in both the x_1 and x_2 directions, then we get two equations

$$\begin{aligned} D_1 v &= 0, \\ D_2 v &= 0. \end{aligned} \tag{2.5}$$

This is an overdetermined system: there are twice as many equations as there are unknown functions. In general, there are no solutions except $v = 0$. To have non-trivial solutions, the connection must satisfy a consistency condition obtained by cross-differentiating (2.5):

$$D_1 D_2 v - D_2 D_1 v = F_{12} v = 0, \tag{2.6}$$

where

$$F_{ab} = [D_a, D_b] = \partial_a A_b - \partial_b A_a + [A_a, A_b] \tag{2.7}$$

is a $\mathfrak{g}$-valued function of x called the *curvature* or *gauge field*. In order for (2.5) to have non-trivial solutions it is necessary and sufficient to have vanishing curvature $F_{12} = 0$.

So the *linear* system (2.5) is consistent if and only if $F_{12} = 0$, which is a *non-linear partial differential equation* for A_a. But this non-linear PDE is rather trivial: up to a gauge transformation, the solution of $F_{12} = 0$ is $A_1 = A_2 = 0$; or to put it differently, the general solution is the pure gauge $A_a = \Lambda^{-1}\partial_a\Lambda$, where $\Lambda(x)$ is arbitrary. How can one get something non-trivial? The crucial idea is to introduce a *parameter*, sometimes called the spectral parameter. In Nigel Hitchin's lectures this is denoted z, in Graeme Segal's it is called λ.

3 First example: self-dual Yang–Mills

Our aim is to define a linear system similar to (2.5), but involving a parameter λ. A straightforward possibility for the parametric dependence is that it be polynomial in λ. In fact, to begin with, we will concentrate on the case of linear dependence (generalizations of this simplest case will be mentioned later). The basic structure we work with is a rank p vector bundle over $\mathbb{R}^4$, and the linear system is obtained from (2.5) by replacing D_1 by $D_1 + \lambda D_3$ and D_2 by $D_2 + \lambda D_4$:

$$\begin{aligned} (D_1 + \lambda D_3)v &= 0, \\ (D_2 + \lambda D_4)v &= 0. \end{aligned} \tag{3.1}$$

Once again, (3.1) is an overdetermined linear system; and it is consistent if and only if

$$[D_1 + \lambda D_3, D_2 + \lambda D_4] = 0. \tag{3.2}$$

This curvature is a quadratic polynomial in λ, so demanding consistency for all λ gives three (matrix) equations for the four (matrix) functions A_a $(a = 1, 2, 3, 4)$, namely

$$\begin{aligned} F_{12} &= 0, \\ F_{14} + F_{32} &= 0, \\ F_{34} &= 0, \end{aligned} \tag{3.3}$$

which are called the *self-dual Yang–Mills (SDYM)* equations. This is an underdetermined system (three equations for four functions), reflecting the gauge invariance of the system. Note that the curvature transforms under a gauge change as

$$F_{ab} \mapsto \widehat{F}_{ab} = \Lambda F_{ab} \Lambda^{-1}, \tag{3.4}$$

so that vanishing curvature is a gauge-invariant condition. Choosing a gauge introduces an extra equation which makes up the deficit.

Why is it called self-dual Yang–Mills? Suppose that $\mathbb{R}^4$ is equipped with the pseudometric $\mathrm{d}s^2 = \mathrm{d}x^1\mathrm{d}x^4 - \mathrm{d}x^2\mathrm{d}x^3 = \eta_{ab}\mathrm{d}x^a\mathrm{d}x^b$, which has signature $++--$. Then there is an associated duality operator (the Hodge dual) which maps k-forms to $(4-k)$-forms, and in particular 2-forms to 2-forms:

$$\omega_{ab} \mapsto (*\omega)_{ab} = \frac{1}{2}\eta_{ac}\eta_{bd}\epsilon^{cdij}\omega_{ij}, \tag{3.5}$$

where ϵ^{cdij} is the totally antisymmetric tensor with $\epsilon^{1234} = -1$. The SDYM equations then say that $*F = F$; in other words, F is self-dual. These equations are invariant under $SO(2,2)$, that is, linear transformations of $\mathbb{R}^4$ which preserve the pseudometric $\mathrm{d}s^2$ (in fact, also under the conformal group in $2+2$ dimensions).

This explains the 'self-dual' part of the name. The Yang–Mills equation is the Euler–Lagrange equation of the Lagrangian density

$$\mathcal{L} = \mathrm{tr}(F_{ab}F_{cd})\eta^{ac}\eta^{bd}, \tag{3.6}$$

namely $\eta^{ca}D_cF_{ab} = 0$. The action of the covariant derivative on the curvature is defined as

$$D_cF_{ab} = \partial_cF_{ab} + [A_a, F_{ab}]; \tag{3.7}$$

note that D_cF_{ab} transforms under gauge change in the same way as F_{ab}. Since the Bianchi identity $\eta^{ca}D_c * F_{ab} = 0$ is automatically true for any connection, it is clear that any connection with self-dual curvature satisfies the Yang–Mills equation (of course, the converse is false). While the Yang–Mills equations are not integrable (in fact, they admit chaotic solutions), the SDYM equations are integrable, and structures can be set up which enable one to generate lots of solutions. These structures naturally lead to *twistor space.* Before describing this, I want to make a few remarks, and describe a more direct solution-generating method.

- One can similarly define the SDYM equations on Euclidean $\mathbb{R}^4$, and these are invariant under $SO(4)$. One needs a mixed metric signature to introduce 'time' into dimensional reductions of the system. In both 4 and 2+2 dimensions the Hodge dual satisfies $** =$ identity, so that $*$ has eigenvalues ± 1, and real self-dual curvatures can exist. In $3+1$ dimensions, however, $** =$

−identity, so $*$ has eigenvalues $\pm \mathrm{i}$, and the SDYM equations have no real solutions.

- We have $4p^2$ functions on $\mathbb{R}^4$ (the matrix components of each A_a), and they satisfy the coupled non-linear PDEs $*F = F$, where $F_{ab} = \partial_a A_b - \partial_b A_a + [A_a, A_b]$. If $p = 1$, then the equations are linear; so to get something interesting we need to take $p \geq 2$.
- Many (in fact most) well-known integrable systems (for example, the static monopoles referred to earlier, Korteweg–de Vries, non-linear Schrödinger, sine-Gordon equation, etc.) are *reductions* of SDYM: see Mason and Woodhouse (1996).

As an example of how one can generate solutions of SDYM, consider the case where $p = 2$ and the gauge group is $SU(2)$, so A_a takes values in $su(2)$. Suppose we had two independent solutions v_1 and v_2 of the linear system (3.1); arrange these two column vectors side-by-side to form a 2×2 invertible matrix $\psi(x, \lambda)$ satisfying

$$\begin{aligned} (D_1 + \lambda D_3)\psi &= 0, \\ (D_2 + \lambda D_4)\psi &= 0. \end{aligned} \tag{3.8}$$

Recalling that $D_a\psi = \partial_a\psi + A_a\psi$, we see that these give

$$\begin{aligned} A_1 + \lambda A_3 &= (\partial_1\psi + \lambda\partial_3\psi)\psi^{-1}, \\ A_2 + \lambda A_4 &= (\partial_2\psi + \lambda\partial_4\psi)\psi^{-1}; \end{aligned} \tag{3.9}$$

so if we know ψ, then we know the right-hand sides for all λ, and we can reconstruct A_a. These will automatically be solutions of SDYM.

But this supposes that we know ψ in the first place. The trick is to assume that ψ has an appropriate simple form, in particular that the λ-dependence is simple. For example, let us assume that ψ is rational in λ, with simple poles at the constant complex numbers $\mu_1, \mu_2, \ldots, \mu_n$ (constant meaning independent of x^a). Since the gauge group is $SU(2)$, ψ has determinant 1 (or at least $\det\psi$ is constant), and ψ is unitary for real λ. In fact, the unitarity condition for general λ is

$$\psi(x, \lambda)\psi(x, \bar{\lambda})^* = I, \tag{3.10}$$

where $*$ denotes the complex conjugate transpose of the matrix. Assume that ψ has the form

$$\psi = I + \sum_{k=1}^{n} \frac{M_k(x)}{\lambda - \mu_k}. \tag{3.11}$$

Then impose unitarity: the vanishing of the residues at the poles on the left hand side of (3.10) implies that M_k has rank 1: $M_k = u_k \otimes v_k$, with a formula for u_k in terms of v_k (or vice versa). One can then substitute this into (3.9), and demand that the right-hand sides be linear in λ. This requirement is satisfied if v_k is an arbitrary function of the combinations $\mu_k x^1 - x^3$ and $\mu_k x^2 - x^4$.

In this way, we get large families of solutions in terms of rational expressions involving arbitrary functions. Typically these solutions represent n-soliton solutions. In the simplest case $n = 1$, we get

$$\psi(x, \lambda) = I + \frac{\mu - \bar{\mu}}{\lambda - \mu} \left(\frac{v^* \otimes v}{v \cdot v^*} \right). \tag{3.12}$$

To end this section, and lead into the next, we note that the two combinations above are the evaluation at $\lambda = \mu_k$ of

$$\omega^1 = \lambda x^1 - x^3, \qquad \omega^2 = \lambda x^2 - x^4. \tag{3.13}$$

Each of these ω^j is annihilated by both the operators

$$\begin{aligned} \eth_1 &= \partial_1 + \lambda \partial_3, \\ \eth_2 &= \partial_2 + \lambda \partial_4, \end{aligned} \tag{3.14}$$

which are the derivative operators in the linear system (3.1). This can be viewed as coming from the underlying twistor geometry.

4 Twistor space and holomorphic vector bundles

For more details on the material of this section, see Ward and Wells (1990). Think of the x^a as being four complex variables, so that $M = \mathbb{C}^4$. In addition, we will allow λ to take all values in the extended complex plane $\mathbb{C} \cup \{\infty\} = \mathbb{P}^1$. An object like $v(x, \lambda)$ is a function on $F = M \times \mathbb{P}^1$, a five-dimensional complex manifold. The first order differential operators $\eth_i$ correspond to vector fields on F, and these span complex planes in F. The space of these planes, i.e. the quotient of F by the distribution $\{\eth_1, \eth_2\}$, is a three-dimensional complex manifold called *twistor space* T. As local coordinates on T, we can use combinations of $\lambda, x^1, x^2, x^3, x^4$ which are annihilated by both $\eth_1$ and $\eth_2$, namely λ, $\omega^1 = \lambda x^1 - x^3$ and $\omega^2 = \lambda x^2 - x^4$. The point in T corresponding to a plane $\widetilde{Z}$ will be denoted Z.

The linear system (3.1), provided the connection is self-dual, gives the condition for p-vector fields (sections of a rank p vector bundle) to be covariantly constant in the $\eth_1$ and $\eth_2$ directions, that is, over the planes spanned by $\eth_1$ and $\eth_2$. So the space of covariantly constant p-vector fields on one of these planes $\widetilde{Z}$ is a p-dimensional vector space. This gives an allocation of a p-dimensional vector space E_Z to each point Z of twistor space, in other words, a rank p holomorphic vector bundle over T. The important fact is that this vector bundle contains all the information of the original connection on M. On M one has a trivial bundle V (which contains no information apart from the integer p), and a connection satisfying some non-linear PDEs; on T, one has just a vector bundle (no connection), and no differential equation at all (cf. the Fourier and inverse scattering transforms, where the differential equation is transformed away). These two structures are equivalent: one can transform back and forth between them. The transformation from M to T is described above. To go the other way requires a bit more work; we shall do this in the next section for a special reduced case, where all dimensions are reduced by 1. The reduction is obtained by demanding that all functions on M are independent of x^1+x^4; or equivalently that they depend on x^1 and x^4 only through the combination x^1-x^4. This effectively reduces M to a three dimensional space-time with signature $++-$. The corresponding reduced twistor space is sometimes called 'minitwistor space'.

5 Yang–Mills–Higgs solitons and minitwistor space

What happens to the SDYM equations when we reduce from $\mathbb{R}^4$ to $\mathbb{R}^3$ as described above? One still has four matrix-valued functions A_a, but one of them (or rather the combination $A_1 + A_4$) is not a connection coefficient, since the corresponding dimension no longer exists. It is now referred to as a Higgs field, and denoted Φ.

For the sake of neatness, let us rename the space-time coordinates: we're on $\mathbb{R}^3$ with coordinates $X^\mu = (X^0, X^1, X^2) \equiv (t, x, y)$, and pseudometric $\mathrm{d}s^2 = \eta_{\mu\nu}\mathrm{d}X^\mu \mathrm{d}X^\nu$, where

$$(\eta_{\mu\nu}) = \begin{pmatrix} -1 & 0 & 0 \\ 0 & 1 & 0 \\ 0 & 0 & 1 \end{pmatrix}. \qquad (5.1)$$

The Hodge duality operator $*$ now maps 2-forms to 1-forms, and the SDYM equations reduce to

$$D\Phi = *F, \tag{5.2}$$

where $D_\mu \Phi = \partial_\mu \Phi + [A_\mu, \Phi]$. The system (5.2) is a set of coupled non-linear PDEs for (A_μ, Φ). (The corresponding Euclidean equations are called the Bogomol'nyi equations for static monopoles.) The linear system which has (5.2) as its consistency condition is

$$\begin{aligned} (\lambda D_x - D_t - D_y + \lambda\Phi)v &= 0, \\ (\lambda D_t - \lambda D_y - D_x + \Phi)v &= 0. \end{aligned} \tag{5.3}$$

We can generate solutions of (5.2) directly, by assuming that we have two independent solutions of (5.3) forming a matrix ψ of the form

$$\psi = I + \sum_{k=1}^{n} \frac{M_k(X)}{\lambda - \mu_k} \tag{5.4}$$

(details as in Section 3). Here M_k is determined in terms of a vector v_k which depends on X^μ only through the combination

$$\omega_k = (t + y)\mu_k^2 + 2x\mu_k + (t - y). \tag{5.5}$$

Of most interest are the solutions which are smooth everywhere in space-time, and localized in space. This requires that the v_k be rational functions of ω_k. The picture that one gets is of n solitons moving in the (x, y) plane. The k-th soliton moves with a constant velocity determined by μ_k. Its shape is determined by the rational function of v_k referred to above, and generically it looks like d_k separate lumps (drifting along at the same velocity), where d_k is the degree of the rational function.

The simplest example is $n = 1$, $\mu = i$, $d = 1$. Note that in this case

$$\begin{aligned} \omega &= -(t + y) + 2\mathrm{i}x + (t - y) \\ &= 2\mathrm{i}(x + \mathrm{i}y); \end{aligned} \tag{5.6}$$

so ω is independent of t ($\mu = \pm i$ give static solutions, all other values of μ do not). We must still choose a unit-degree rational function. One simple possibility is to take $v = (1, \omega)$. The corresponding soliton is a single lump centred at $x = y = 0$. In order to get a picture, it is useful to plot the positive gauge-invariant quantity $-\mathrm{tr}(\Phi^2)$. In this case,

$$-\mathrm{tr}(\Phi^2) = \frac{\kappa}{(1 + x^2 + y^2)^2}, \tag{5.7}$$

where κ is some constant.

If one allows ψ to have higher-order poles in λ (order ≥ 2), then explicit solutions with more complicated behaviour can be found. For example, taking ψ to have a double pole at $\lambda = \mathrm{i}$, and no other poles, gives a solution in which the 'centre of mass' is static, but two solitons collide head-on and emerge at right angles to their line of approach. This type of scattering is familiar for topological solitons (but there, by contrast, one does not have explicit solutions).

It is possible to relate all this to the algebraic geometry of holomorphic vector bundles over minitwistor space, as follows. Complexify $\mathbb{R}^3$ to $\mathbb{C}^3$, with complex metric as before. The twistor space T is the space of *null planes* in $\mathbb{C}^3$ with respect to this metric. A null plane is one of the form $\eta_{\mu\nu}k^\mu X^\nu = \omega$, where ω is constant and k^μ is a null vector:

$$\eta_{\mu\nu}k^\mu k^\nu = -(k^0)^2 + (k^1)^2 + (k^2)^2 = 0. \tag{5.8}$$

One can parametrize such null vectors using a single complex parameter λ, namely

$$k_\mu = \eta_{\mu\nu}k^\nu = (\lambda^2 + 1, 2\lambda, \lambda^2 - 1). \tag{5.9}$$

Since the direction of k_μ is of importance (not its length), one can allow λ to take the value ∞ in this formula, that is, $\lambda \in \mathbb{P}^1$. So (λ, ω) are local coordinates on T. Note that the null plane (λ, ω) is given by

$$\begin{aligned} & (\lambda^2 + 1)X^0 + 2\lambda X^1 + (\lambda^2 - 1)X^2 = \omega \\ \Rightarrow \quad & \omega = (t + y)\lambda^2 + 2x\lambda + (t - y), \end{aligned} \tag{5.10}$$

which when evaluated at $\lambda = \mu_k$ is precisely the combination encountered earlier (5.5). Note also that the two vectors from the linear system (5.3)

$$\begin{aligned} \eth_1 &= \lambda\partial_x - \partial_t - \partial_y \\ \eth_2 &= \lambda\partial_t - \lambda\partial_y - \partial_x \end{aligned} \tag{5.11}$$

are the tangent vectors to the null planes labelled by λ, since they annihilate the linear combination $k_\mu X^\mu$. So, as before, one can interpret T as the quotient space of $\mathbb{C}^3 \times \mathbb{P}^1$ by the distribution $\{\eth_1, \eth_2\}$.

Globally, T is a holomorphic line bundle over $\mathbb{P}^1$. In fact, it is the line bundle whose global holomorphic sections are quadratic polynomials as in (5.10); it follows that T has Chern class 2. By definition, a point in T corresponds to a (null) plane in M. What does a point in M correspond to in T? Fix $p = (t, x, y) \in M$. Then (5.10) gives ω as a function of λ, or more precisely, a global holomorphic section of the line bundle T. Geometrically, this is a curve in T, and the curve corresponding to $p \in M$ will be denoted $\widehat{p}$. In summary, there is a 'duality' between M and T:

$$\begin{aligned} \text{point in } M &\longrightarrow \text{curve in } T \\ \text{plane in } M &\longleftarrow \text{point in } T. \end{aligned}$$

Recall that if we start with a Yang–Mills–Higgs field (A_μ, Φ) satisfying $D\Phi = *F$, then (5.3) can be integrated: the space of vector fields covariantly constant on a null plane $\widetilde{Z}$ is a vector space V_Z, the fibre of a holomorphic vector bundle over T.

Theorem 5.1. (Ward 1990) *There is a one-to-one correspondence between:*

(i) solutions (A_μ, Φ) of the equation $D\Phi = *F$ on M [with gauge group $SU(2)$]; and

(ii) holomorphic [$SU(2)$] vector bundles V over T satisfying the triviality condition

$$V|_{\widehat{p}} \text{ is trivial } \forall p \in M. \tag{5.12}$$

[The $SU(2)$ part is inessential, and we will ignore it.]

It is clear why we get the triviality property (5.12): any two points Z, Y on $\widehat{p}$ correspond to null planes $\widetilde{Z}, \widetilde{Y}$ through the single point $p \in M$. This gives a natural identification of the fibres $V_Z \cong V_Y$, namely a covariantly constant vector field on $\widetilde{Z}$ will be identified with a covariantly constant vector field on $\widetilde{Y}$ if the two fields coincide at the common point p. This defines a trivialization of $V|_{\widehat{p}}$.

We have described previously how to get from (i) to (ii) (see Section 4). It remains to describe how to get from (ii) to (i). Given V

satisfying (5.12), the holomorphic sections of $V|_{\widehat{p}}$ form a (2-dimensional) vector space, which defines the fibre E_p of a vector bundle E over M (E is necessarily trivial, since $M = \mathbb{C}^3$). Note that *compactness* of T in the λ coordinate is essential for this: if T were a line bundle over $\mathbb{C}$ rather than the compact Riemann surface $\mathbb{P}^1$, then $\widehat{p}$ would be non-compact and its holomorphic sections would form an *infinite*-dimensional vector space. So allowing the spectral parameter λ to take the value ∞ is crucial. We now need to reconstruct (A_μ, Φ). If p and q are null-separated (lie on the same null plane $\widetilde{Z}$), then the curves $\widehat{p}$ and $\widehat{q}$ intersect at a single point Z. This allows us to identify holomorphic sections of $V|_{\widehat{p}}$ with those of $V|_{\widehat{q}}$ (namely, two such sections are identified if they coincide at Z), and hence define an identification of fibres E_p and E_q where p and q are null-separated. So given a vector $v \in E_p$ one can propagate it 'constantly' over any null plane $\widetilde{Z}$ through p, by defining its value at $q \in \widetilde{Z}$ to be that holomorphic section of $V|_{\widehat{q}}$ whose value at Z coincides with v. The coefficients appearing in the propagation law are identified as (A_μ, Φ).

Since any solution of the Yang–Mills–Higgs equations corresponds to a holomorphic vector bundle over T, one might ask whether it is possible to parametrize the space of such bundles. If so, all solutions of the equations would, in some sense, be 'known'. In fact, since T is non-compact (being the total space of a line bundle), it is impossible to parametrize the moduli space of vector bundles over T explicitly. However, for pure soliton solutions, such as those described earlier, the vector bundle V extends to a compactification $\overline{T}$ of minitwistor space: each fibre L_λ of T is compactified from $\mathbb{C}$ to $\mathbb{P}^1$. Now $\overline{T}$ is an algebraic variety, and bundles can be constructed by specifying rational data (cf. the rational functions in the soliton solutions described previously); see Ward (1990).

6 Generalizations

Recall that the SDYM equations arose from an overdetermined linear system consisting of two equations, each polynomial of degree one in one parameter λ. Each equation had the form

$$L^a(\lambda)(\partial_a + A_a(x))v(x, \lambda) = 0, \tag{6.1}$$

where the L^a are four linear polynomials in λ, and the A_a are matrices. How can one generalize this? There are some obvious possibilities:

(A) one can allow the functions $L^a(\lambda)$ to depend on x;

(B) one can increase the values of any or all of the integers mentioned above: for example, by allowing more than one parameter, or polynomials of higher degree, or more than two linear equations in the system;

(C) one can use more general linear operators, for example $L^a(\lambda)$ could be matrices (this includes the possibility of higher-order differential operators);

(D) one could replace the differential operators by difference, integral or pseudo-differential operators.

All of these lead to interesting integrable systems, some of which are as follows.

(A) Put $A_a(x) = 0$, and allow L^a to depend on x. Then the coefficients of $L^a(x, \lambda)$ become the dynamical fields. There is no longer a gauge field, but the underlying space M becomes curved, and its metric satisfies the self-dual Einstein equations. On the four-dimensional manifold M, we have two vector fields, depending linearly on λ,

$$\begin{aligned} \eth_1 &= V_1 + \lambda V_3 \\ \eth_2 &= V_2 + \lambda V_4. \end{aligned} \tag{6.2}$$

The consistency condition for the linear system $\eth_1 v = 0 = \eth_2 v$ is that these two vector fields should commute. Then $\eth_1$, $\eth_2$ span surfaces in M. The space of these surfaces forms a three-dimensional complex manifold T, which is a deformed version of the flat twistor space encountered before. Like the original T, it is fibred over $\mathbb{P}^1$. The metric on M is defined (up to conformal scale) by demanding that $\eth_1$, $\eth_2$ are null vectors for all λ. The conformal curvature of this metric is self-dual, and this leads to the self-dual Einstein equations (some extra structure is required to fix the conformal scale). The crucial point is that given T, we can reconstruct the geometry of M. Namely, the sections $\widehat{p}$ of T, thought of as a bundle over $\mathbb{P}^1$, are by definition the points p of M. The set of all sections $\widehat{p}$ through $Z \in T$ gives a surface $\widetilde{Z}$ in M. Defining such surfaces to be totally null defines the conformal structure on M. For more details, see Ward and Wells (1990).

(B) Increasing the number of equations in the linear system from 2 to $2k$ gives a generalization of SDYM to $4k$ dimensions (Ward 1984).

(AB) The self-dual Einstein equations generalize to $4k$-dimensional hyperkähler geometry, of interest to geometers, and also occurring in moduli spaces of topological solitons (and elsewhere in mathematical physics).

(C) Allowing $L^a(\lambda)$ to be matrix-valued, one can obtain the Davey–Stewartson and KP equations (well-known three-dimensional PDEs in soliton theory).

(D) Using difference operators leads to integrable lattice systems and cellular automata.

To a geometer, generalizations (A) and (B) are natural, because first-order scalar (i.e. not matrix) differential operators $L^a \partial_a$ can be interpreted as vector fields on M. Taking the quotient of M by these vector fields leads to twistor space. By contrast, with generalizations (C) and (D) one loses this nice geometry. There are current efforts to find a generalized geometry which can include these latter two cases, but whether these will be useful remains to be seen.

Bibliography

Mason, L. J. and Woodhouse, N. M. J. (1996). *Integrability, self-duality, and twistor theory.* Oxford University Press.

Ward, R. S. (1984). Completely solvable gauge-field equations in dimension greater than four. *Nucl. Phys.*, B**236**, 381–96.

Ward, R. S. (1990). Classical solutions of the chiral model, unitons, and holomorphic vector bundles. *Commun. Math. Phys.*, **128**, 319–32.

Ward, R. S. and Wells, R. O. (1990). *Twistor Geometry and Field Theory.* Cambridge University Press.

Index

9 780198 504214

Milton Keynes UK
Ingram Content Group UK Ltd.
UKHW050446010724
444910UK00005B/52